BIO-BASED POLYMERS
PROPERTIES AND APPLICATIONS

BIO-BASED POLYMERS
PROPERTIES AND APPLICATIONS

Grace Potter
Sharon Hunt

Kruger Brentt
Publishers

2024

Kruger Brentt Publishers UK. LTD.
Company Number 9728962

Regd. Office: 68 St Margarets Road, Edgware, Middlesex HA8 9UU

© 2024 AUTHORS
ISBN: 978-1-78715-140-6

For information on all our publications visit our website at http://krugerbrentt.com/

PREFACE

Bio-based polymers are a class of polymers that are derived from renewable biomass sources such as plants, algae, and microorganisms. They are becoming increasingly popular due to their potential to replace petroleum-based polymers, which are non-renewable and contribute to environmental pollution. In this book, we will explore the properties and applications of bio-based polymers and their potential to revolutionize the field of materials science.

Bio-based polymers have a range of properties that make them attractive for use in various applications. They can be made into flexible or rigid materials and can have a range of mechanical properties, including strength and stiffness. They can also be made into biodegradable or compostable materials, making them a more sustainable alternative to petroleum-based polymers.

One of the most well-known bio-based polymers is polylactic acid (PLA). PLA is derived from corn starch or sugarcane and can be used to make a range of products, including packaging materials, disposable tableware, and textiles. It has similar mechanical properties to traditional petroleum-based polymers and can be composted at the end of its life cycle, making it a more sustainable option.

Another bio-based polymer is polyhydroxyalkanoate (PHA), which is produced by microorganisms such as bacteria. PHA has similar properties to traditional petroleum-based polymers and can be used in a range of applications, including packaging materials and medical devices. It is also biodegradable, making it a more sustainable option than petroleum-based polymers.

Bio-based polymers also have the potential to replace petroleum-based polymers in the production of plastic products. Plastic products are a major contributor to environmental pollution, with a large proportion of them ending up in landfills or the ocean. Bio-based polymers can offer a more sustainable alternative, with some bio-based polymers being biodegradable or compostable. They can also be recycled, reducing the need for new petroleum-based polymers to be produced.

However, there are also challenges associated with the production and use of bio-based polymers. One challenge is the cost of production, which can be higher than petroleum-based polymers. There are also concerns about the availability of biomass sources and the potential impact of large-scale production on land use and food security.

The present book contains 26 chapters namely Introduction; biopolymers: definition, classification and applications; bio-polymers from agricultural waste; bio-based thermosetting resins composed of cardanol novolac and bismaleimide; castor oil as a potential renewable resource for the production of bio based polymer; epoxidized soybean oil (eso)/poly(lactic acid) composites; soybean oils for use as biobased thermoset resins; bio-renewable polymer composites from tall oil-based polyamide and lignin-cellulose fiber; linseed oil-based thermosets; vegetable oil-derived epoxy monomers; lignin-styrene-butyl acrylate based composite; biocomposites from lactic acid thermoset resins; biomaterials from starch; vegetable oil-based bio-polymers: synthesis and applications; renewable bio-polymers from tung oil and natural terpenes; triglycerides and phenolic compounds of bio-based thermosetting epoxy resins; preparation and characterization of a soy protein based bio-adhesive; poly (lactic acid) filled with cassava starch-g-soybean oil maleate; poly (l- lactic acid) filled with cassava starch-g-soybean oil maleate; new bio-composites containing industrial lignins ; synthesis of biocomposites from natural oils; nano-lignin filled natural rubber composites; hybrid bio-based polymers from blends of epoxy and soybean oil resins reinforced with jute woven fabrics; bipolymers from rosin-based chemicals & shellac based polymer. This book highlights the various types of polymers that can be derived from bio- renewable resources. The book addresses bio-based feedstocks, production processes, packaging types, recent trends in packaging, and legislative demands for food contact packaging materials. It explores opportunities for biopolymers in key end-use sectors, the penetration of biopolymer based concepts in the packaging market. This is an excellent book for scientists, engineers, graduate students and industrial researchers in the field of bio-based materials.

We are grateful to all those persons as well as various books, manuals, periodicals, magazines, journals etc. that helped in the preparation of this book. In spite of the best efforts, it is possible that some errors may have occurred into the compilation and editing of the book. Further queries, constructive suggestions and criticisms for the improvement of the book are always welcomed and shall be thankfully acknowledged.

Grace Potter
Sharon Hunt

CONTENTS

1
Chapter

INTRODUCTION

The increasing environmental awareness of the society has become an important factor in recent decades affecting legislation, commerce and industry as well as research and development to a great extent. The same trend can be observed in polymer industry as well, as the production and use of biopolymers increases continuously with a very high rate thus all information on these materials is very important. Biopolymers often have inferior properties compared to commodity polymers. Modification is a way to improve properties and achieve property combinations required for specific applications. One technique is the incorporation of fillers and reinforcements, thus providing various biocomposites. Most studies focus on the potential use of natural lignocellulosic fibers, i.e. wood flour, sisal, flax, etc. into both conventional and biopolymers, although using conventional, mineral fillers is also of high importance. Interfacial interactions play a crucial role in the determination of composite properties, affecting structure and micromechanical deformation processes to a great extent, thus investigation and modification of these is in the focus of scientific interest.

Biopolymers can also be modified by blending which allows considerable improvement in the impact resistance of brittle polymers. However, further study is needed on the miscibility-structure-property relationships of these materials to utilize all potentials of the approach. If possible, interactions are even more important in blends than in other heterogeneous polymeric materials since they determine the mutual solubility of the phases, the thickness and properties of the interphase formed during blending and the structure of the blend. As a consequence, the proper, and possibly quantitative, characterization of interaction is of utmost importance for the prediction of blend properties.

In spite of the polar character of biopolymers, often compatibilization is needed to achieve the properties required for a specific application. Different strategies include various reactive and non-reactive approaches. In the case of non-reactive methods, mostly amphiphilic compounds, like block-copolymers are used for the compatibilization of blends, while surface modification of the filler is a common way of altering interfacial adhesion in polymer composites. The chemical structure of biopolymers, however, opens up possibilities to their reactive modification. Copolymerization, grafting, transesterification, the use of reactive coupling agents have all been utilized with success to achieve polymers, blends and composites with improved properties.

2
Chapter

BIOPOLYMERS: DEFINITION, CLASSIFICATION AND APPLICATIONS

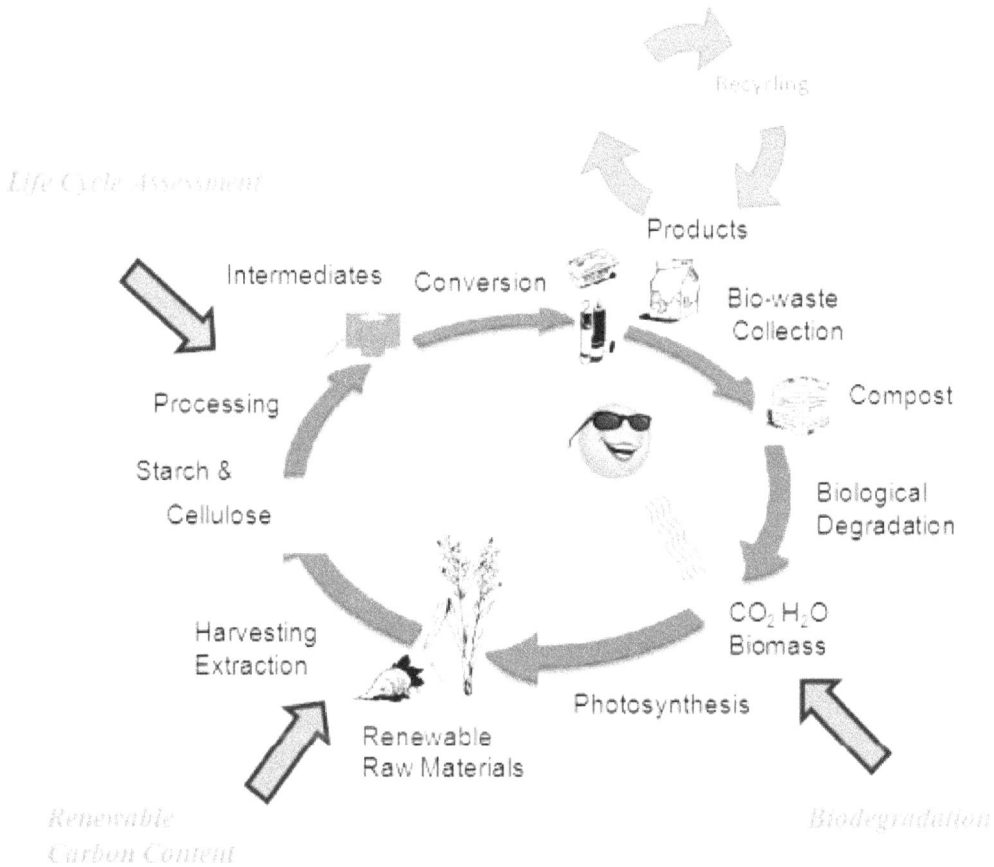

During recent years significant advances have been made in using and development of biodegradable polymeric materials for life applications. Degradable polymeric biomaterials are preferred because these materials have specific physical, chemical, biological, biomechanical and degradation properties. Wide ranges of natural or synthetic biopolymers capable of undergoing degradation hydrolytically or enzymatically are being investigated for many applications. This review aimed to provide an overview for the importance of biomaterials, produced or degraded naturally, classification and applications.

DEFINITION

Recently, concerted efforts to protect the environment not only by using natural renewable materials environmentally friendly, but also using materials that decompose naturally in the environment was done and increasing rapidly.

One of these materials is biopolymers, a kind of polymers, that produced by living organisms such as alginate and carrageenan which produced naturally occurring anionic polysaccharide isolated from the seaweeds, chitosan found in insects and crustaceans shells of certain other organisms, including many fungi, algae, and yeast . Monomeric units of Biopolymers are sugars, amino acids, and nucleotides. Cellulose, starch and chitin, proteins and peptides, DNA and RNA are all examples of biopolymers.

Biopolymers and their derivatives are varied, plentiful and important for life. They exhibit the characteristics of a wonderful and increasingly important for various applications. For example, these biomaterials can be divided into proteins and poly (amino acids), poly- di- and monosaccharides such as cellulose, starch, fructose and glucose.

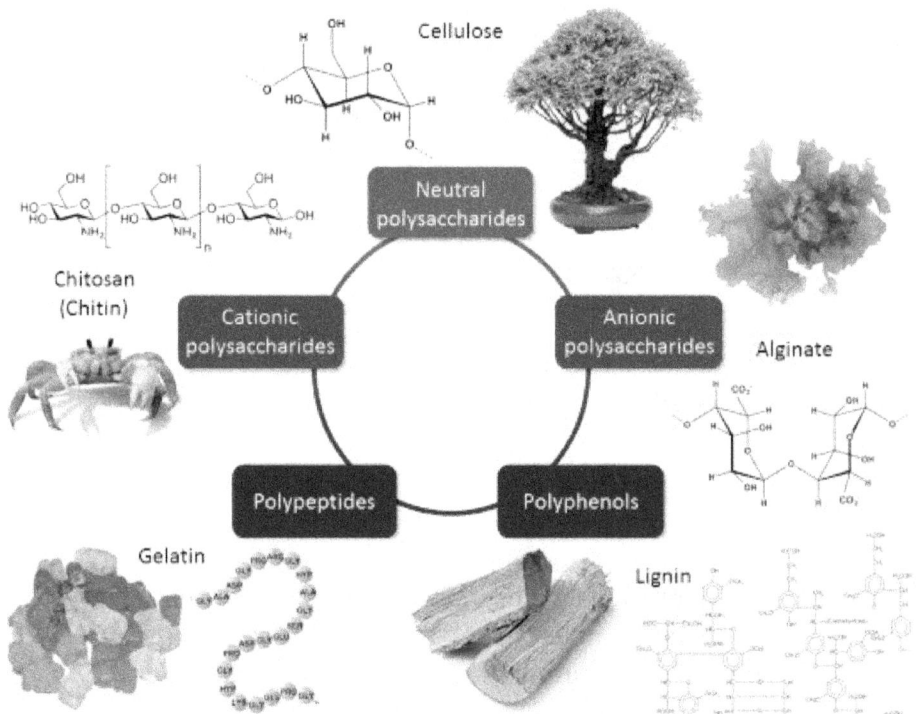

Fig. 1. Natural renewable biomaterials

Fig. 2. Biopolymers in nature

Biodegradable Natural polymers

To select a biomaterial for tissue engineering is the most critical step in scaffolds development. The main requirement should include biocompatible, non-toxic, providing favorable cellular interactions and tissue development, besides, the properties of mechanical strength, tensile strength etc. is also needed of synthetic polymer Biodegradable should be requested and also should support the reconstruction of a new tissue without inflammation. According to the chemical structure of a vast variety of polymers as living organisms synthesized, they can be classified into major classes: (i) polysaccharides, (ii) proteins, and (iii) polyesters. Also basing on the advances in biotechnology, natural polymers can be obtained by the fermentation of microorganisms or in vitro production by enzymatic processes.

Based material classification

Biomaterials can be classified depending on based material; in this case they are classified into four main types of Biopolymers.

Sugar based Biopolymers

Biopolymers based on sugar can be produced by blowing, vacuum forming injection, and extrusion. Lactic acid polymers (Polyactides) are created from milk sugar (lactose) which is extracted from maize, potatoes, wheat and sugar beet. Polyactides, manufactured by methods like vacuum forming, blowing and injection molding, are resistant to water.

Starch based Biopolymers

Biopolymers based on starch acts as natural polymer and can be obtained from vegetables like wheat, tapioca, potatoes and maize.

The material is stored in tissues of plants as carbohydrates. It is composed of glucose and can be obtained by melting starch. This polymer is not present in animal tissues. Dextrans, produced by starch hydrolysis, which is a group of low-molecular-weight carbohydrates, enzymatically synthesized by immobilized Enterococcus faecalis Esawy dextransucrase onto biopolymer carriers.

Cellulose based Biopolymers

Biopolymers based on cellulose are used for packing cigarettes, CDs and confectionary. This polymer is composed of glucose and is the primary obtained from natural resources, plant cellular walls, like cotton, wood, wheat and corn.

Biopolymers based on Synthetic materials

Biopolymers based on Synthetic compounds also used for making biodegradable polymers such as aliphatic aromatic copolyesters are obtained from petroleum. They are compostable and bio-degradable completely though they are manufactured from synthetic components.

BIODEGRADABLE SYNTHETIC POLYMER

Polyglycolide Acid (PGA)

It is a highly crystalline polymer (45–55% crystallinity). It also exhibits a high tensile modulus with a very low solubility in organic solvents. The currently most commonly used in synthetic biodegradable polymer are considered to be poly (glycolic acid) and poly (lactic acid). Polyglycolide was initially investigated for developing resorbable sutures because of its excellent fiber forming ability. Polyglycolides are broken down into glycine or converted into carbon dioxide and water via the citric acid cycle.

Polylactide Acid (PLA)

Similar to polyglycolide, it is also a crystalline polymer (37% crystallinity). It has a glass transition temperature of 60–65°C and a melting temperature of approximately 175°C. Poly (L-lactide) is a slow-degrading polymer compared to polyglycolide. L-lactide (LPLA) homopolymer is a semicrystalline polymer with high tensile strength exhibition, low elongation, and have a high modulus consequently that make them suitable in orthopedic fixation and sutures that have wild applications like load-bearing.

Poly (lactide-co-glycolide)

In adapting the property of PGA in wide application, copolymer of PGA with the more hydrophobic poly (lactic acid) (PLA) were investigated more intensive in Maurus PB`s team review. Compared to homopolymers, the intermediate co-polymers were found to be less stable. Different ratios of poly (lactide-co-glycolides) have been commercially developed and are being investigated for a wide range of biomedical applications.

Polycaprolacton

A semicrystalline polymer with high solubility of polycaprolactone and low melting point (59-64°C) can stimulate its biomaterial application research. Due to the short half-life, there is a problem with delivery of bioactive agent, therefore, a proper transport of the bioactive agents like drug; enzyme etc. a vehicle is needed. Starch-poly-epsilon-caprolactone nmicroparticles then were developed in some applications such as drug delivery and tissue engineering applications. In addition, an emulsion solvent extraction/evaporation technique was prepared in these microparticles.

Poly dioxanone

Poly dioxanone (PDS) is a colorless, non-toxic and semicrystalline polymer. It was the choice material for commercially developed monofilament suture in the 1980s which with a glass transition temperature of -10 to 0°C together with approximately 55% crystallinity. PDS may be broke down into glycoxylate or converted into glycine and subsequently into carbon dioxide and water similar to polyglycolides.

Polyurethane

Polyurethanes are generally prepared by the polycondensation reaction of diisocyanates with alcohols and/amines. Mainly usage for cardiovascular diseases like pacemakers and vascular grafts, bio-stable polyurethanes and poly (ether urethanes) are investigated for the future view in medical implanting extensively. Flexibility and mechanical strength are the reason why the polyurethane is the biocompatible used in a wide range of medical devices. Also for the biocompatibility, polyurethane can attach to membrane and don't stimulate immune reactions easily. A semipermeable membrane is also developed by Knauth team who work in artificial skin for premature neonates.

POLYHYDROXYBUTYRATE (PHB), POLYHYDROXYVALERATE (PHV) AND COPOLYMERS

Both PHB and PHB-PHV can be processed into different shapes and structures i.e. films, sheets, spheres and fibers. In addition, both PHB and PHB-PHV are also found to be soluble in the wide range of solvents. Polymers like polyhydroxybutyrate (PHB), polyhydroxyvalerate (PHV), and copolymers with short biodegradation time are derived from microorganism's fermentation. PHB has high potential in gas barrier and could degrade to D-3 hydroxybutyric acid in vivo together with a low toxicity. Not only controlled drug release, sutures, and artificial skin was adapted but also has been investigated as a material for bone pins and plates development. Bioactive ceramics are also added to make them become potential biopolymer, these bioactive ceramics are hydroxyapatite (HA) and tricalcium phosphate (TCP) that could better enhances the ability of the composites to induce the formation of bone-like apatite on their surfaces.

Polyanhydried

As a possible substitute for polyesters in textile application polyanhydrides were explored and due to its pronounced hydrolytic instability it is still a failor. Owing to this problem, an exploration with polyanhydride as degradable implant material was carried out by Domb`s team. In addition to the benefit of polyanhydride, it can also possess an excellent biocompatibility in vivo. Poly [(carboxyphenoxy propane)-(sebacic acid)] (PCPP-SA) is considered as the most extensively investigated polyanhydride. To compress molding or microencapsulation, drug loaded device has been best prepared and used to deliver the bis-chloroerhylnitrosourea (BCNU) to the brain which can treat brain cancer. Another potential vehicle for gentamicine to treat osteomyelitis was found by a co-polymer of 1:1 sebacic acid and erucic acid dimer. Compare drug delivery and clinical trial, various polymers are available for localized drug delivery but more efficient polymer for drug delivery was shown by polyanhydride.

Poly (ortho esters)

Poly (ortho esters) is a degradable polymer which has become into developing in recent number of years. Poly (ortho esters) erosion in aqueous environments is very slow because it has hydrophobic properties. The surface erosion mechanism is not the only unique feature of poly (ortho esters) but also with the rate of degradation for these polymers, pH sensitivity, and flexibility in the glass transition temperatures which can be controlled by using varying levels of chain diols.

BIOPOLYMERS TYPES

Two different criteria underline the definition of a "biopolymer" (or "bio plastic"): (1) the source of the raw materials and (2) polymer biodegradation.

Biopolymers made from renewable raw materials (bio-based), and being biodegradable

These polymers can be produced by either biological systems (microorganisms, animals, and plants) or synthesized chemically from biological starting materials (e.g., corn, sugar, starch, etc.) Biodegradable bio-based biopolymers include (1) synthetic polymers from renewable resources such as poly (lactic acid) (PLA); (2) biopolymers produced by microorganisms, such as PHAs; (3) natural occurring biopolymers, such as starch or proteins— natural polymers are by definition those which are biosynthesized by various routes in the biosphere. The most used bio-based biodegradable polymers are starch and PHAs.

Biopolymers made from renewable raw materials (bio-based), and not being biodegradable

These biopolymers can be produced from biomass or renewable resources and are non-biodegradable. Non-biodegradable bio-based biopolymers include (1) renewable resource's synthetic polymers such as specific polyamides from castor oil (polyamide 11), specific polyesters based on bio propanediol, bio polyethylene (bio-LDPE, bio-HDPE), bio polypropylene (bio-PP), or bio poly (vinyl chloride) (bio-PVC) based on bioethanol (e.g., from sugarcane), etc.; (2) natural occurring biopolymers such as natural rubber or amber .

Biopolymers made from fossil fuels, and being biodegradable

These biopolymers are produced from fossil fuel, such as synthetic aliphatic polyesters made from crude oil or natural gas, and are certified biodegradable and compostable. PCL, poly (butylene succinate) (PBS), and certain "aliphatic–aromatic" copolyesters are at least partly fossil fuel-based polymers, but they can be degraded by microorganisms.

Shapes

Biopolymers have been classified by the shape into many types. These types such as disk, beads; thin films (membrane) and nanoparticles as follow:-

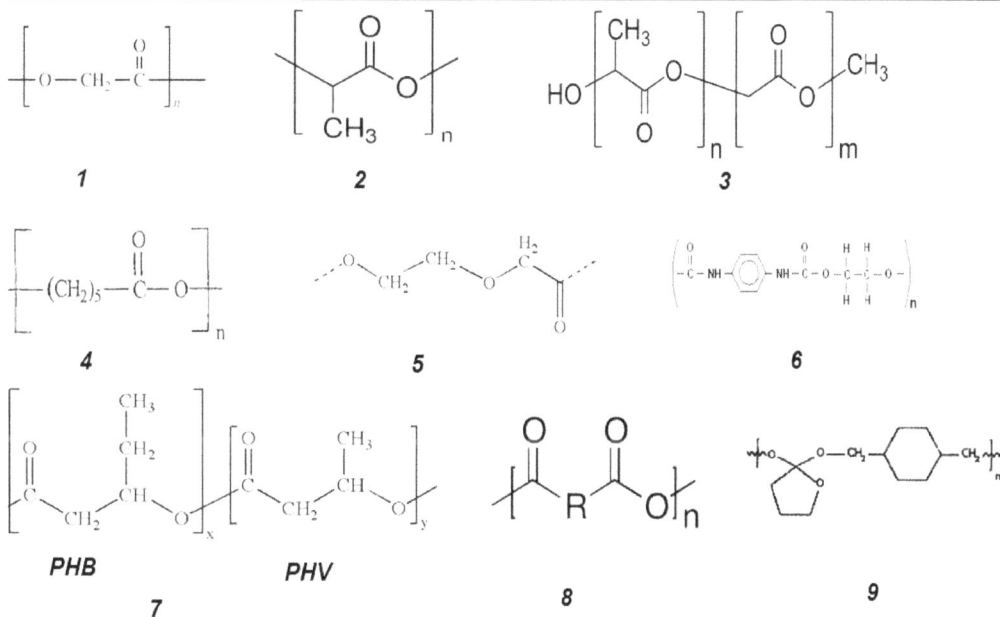

Fig. 3. Biodegradable synthetic polymers; 1: Polyglycolide Acid; 2: Polylactide Acid; 3: Poly (lactide-co-glycolide); 4: Polycaprolacton; 5: Poly dioxanone; 6: Polyurethane; 7: Polyhydroxybutyrate, polyhydroxyvalerate; 8: Polyanhydried; 9: Poly(ortho esters)

Disk

Gel disks are widely used in the literature. Researchers usually use the casting method, e.g. a Petri dish, to make a single film of gel and then cut it into disks using cork borers. Elnashar et al., invented a new equipment to make many uniform films in one step and with high accuracy using the equipment "Parallel Plates" as shown in Fig. 4.

Fig. 4. Parallel plates equipment for making uniform gel disks

Beads

Gel beads are mostly used in industries as they have the largest surface area and can be formed by many techniques such as the interphase technique, ionic gelation methods, dripping method and the Innotech Encapsulator. The Innotech Encapsulator as shown in Fig. 5 has the advantage of high bead production (50 – 3000 beads per second depending on bead size and encapsulation-product mixture viscosity), which is suitable for the scaling up production on the industrial scale.

Fig. 5. Inotech Encapsulator IE-50 R

Membrane

Polymeric membrane can be done by dissolving the polymer material in its solvent and then casting, washing and dryness as in Fig. 6. Sakurai`s research team give observation on the domain structure and show the conclusion that the intracellular concentration of calcium ions, a second messenger in the transmission of biological signals or an enzyme cofactor in the coagulation system, get the influences from polymeric materials. It also reported in the study where the membrane from nonthrombogenicity of a PEUU can be improved remarkably by surface-grafting of polyoxyethylene chains.

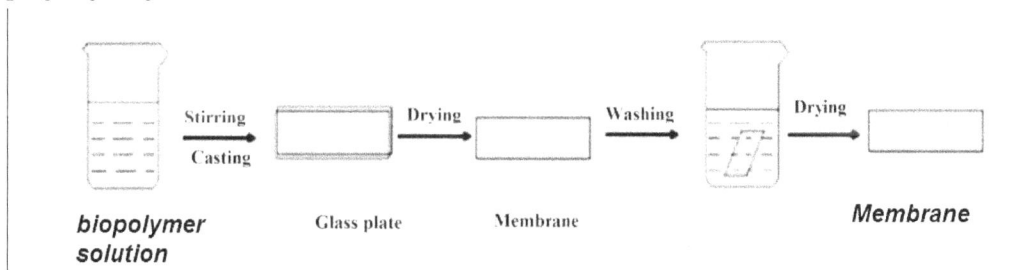

Fig. 6. Membrane preparation method

Nanoparticles

Nanoparticles are defined as sub-micron solid particles, which can be used for nano-encapsulation of bioactive compounds. Nanoparticles, nano-spheres or nano-capsules can be obtained depending on the method of preparation. Nano-capsules are vesicular systems in which the bioactive compounds are confined to a cavity consisting of an inner liquid core surrounded by a polymeric membrane while Nano-spheres are matrix systems in which the bioactive compounds are physically and uniformly dispersed, biodegradable polymeric nanoparticles can be produced from proteins (such as, gelatin and milk proteins), polysaccharides (such as chitosan, sodium alginate and starch) , and synthetic polymers (such as poly (D, Llactide), poly (lactic acid) PLA , poly (D, L-glycolide), PLG, poly (lactide-co-glycolide), PLGA, and poly (cyanoacrylate) PCA).

Recently, different methods have been adopted for preparation of nanoparticles from natural macromolecules (like chitosan, sodium alginate, gelatin, etc). Konecsni and Nickerson prepared chitosan nanoparticles containing rutein by using ionic gelation method.

Applications

Because of the great importance of these biomaterials, it has many important uses that overlap in various important applications in our life.

Industrial applications

Biopolymers used in industry to save money and time especially in immobilization techniques, for example penicillin G Acylase was immobilized onto poly vinyl chloride to produce 6- amino peicillinc acid which is the back bone in b- lactam antibiotic industry. Alginate, carrageenan and carboxy methyle cellulose was used to encapsulate date Palme extract and levan immobilization.

ACCELERATION OF ENDOTHELIALIZATION

Herring`s team synthesized a cell-hybridization-type artificial vascular tissue successfully. But the survival of collected cells is very difficult in seeding autogolal vascular endothelial cells to fibrin layer formed on a polymer membrane by preclotting (seeded graft). Avoiding the problem with this difficulty, cultured graft was proposed, in which endothelial cells are adhered and grown up in advance on the surface of porous membrane to cover it completely with a layer of endothelium but a steady supply of living cells and an immune response comes as the second problems.

Polymer membranes is strongly desired and to obtain cell-hybridization materials without encountering these problems this is because of the living cells can be easily adhere and proliferate upon implanting in a living body, the design and synthesis of polymer membranes is also strongly needed where biopolymer has a good future. Chitin in wound healing dressing is used from its ability for N-acetylglucosamine which can not only accelerate the rate of tissue repair but also prevent the formation of scars and contraction of the skin. Using as a powder form, however, chitosan and chitin are now being incorporated in films, membranes, gels and woven or non- woven dressings.

CONJUGATION OF REAGENT SUPPRESSING PLATELET ACTIVATION

Platelets on the material's surface get influence from adhesion and activation works in thrombus formation. Reagents suppress platelet activation, adrenalin-shielding reagent, drugs participating in the prostaglandin metabolic system can affect the cyclic nucleotide production, and those participating in the Ca2+ regulating mechanism are considered. Polymer membranes immobilize platelet tranquilizers to yield highly nonthrombogenic materials. Ebert et al.; report that PGI2-immobilized polystyrene beads has a great inhibition effect on platelet in both adhesion and activation and their team also pointed out that adhesion and activation of platelets are suppressed on heparin-immobilized materials too.

In cosmetic aspect

The non-toxic and environmentally friendly nature of the PLPs's advantange contribute to the invention of cosmetic use. The patent from Dimotakis group claims that the solubility of the PLPs in both aqueous and non-aqueous polar solvents provide versatility where amenable formulate into any type of cosmetic composition for the invention and use of keratinous tissue. And the patent also stimulate that the PLP has the capability in absorbing light outside the visible range and emit light with a good quantum yield. With substantially the same intensity where the light is absorbed, they successfully invent the function of UV-protectant effect and skin enlighting or colorizing effect according to the use of photoluminescent polymers that polymers having grafted thereon or incorporated into the amino acid or an amino acid derivative's backbone or skeleton.

Food Packing material with safety

Nano materials for polymer modification bring a breakthrough in recent years. The advantage of natural or synthetic polymer with food packaging materials, mechanical strength, flexibility, heat resistance, barrier properties, and other properties have been effectively improved. In food packaging, there are starch nano composite materials from which starch / montmorillonite nano composite film, cellulose nano composite materials, protein nano composite materials and poly lactic acid nano composite materials were prepared. As a food packaging material, it can also improve the antimicrobial effect of packaging, mechanical strength, flexibility, heat resistance and barrier properties, etc. Many countries in Europe and the United States give research through the animal in vivo and in vitro experiments then demonstrated that nano particles have a certain degree of toxicity in animal organs, respiratory and cardiovascular systems which will cause certain damage. As a result of this conclusion, the sixth meeting of the official nano materials in EU was held in the Belgian capital Brussels on November 16, 2008 with the aim at "registration, evaluation, authorization and restriction of chemicals" and regulated that industrial raw materials was included in the scope of regulation.

Medical industry

As the first biodegradable sutures was approved in the 1960s. Polymers have vertical uses in medical industry from the preparation of polycaprolactone and glycolic acid. Multy purpose of products basing on glycolic acid, lactic acid and other materials, including poly (dioxanone), poly(trimethylene carbonate) copolymers, and poly(-caprolactone) homopolymers and copolymers have been wildly used as medical devices. Accompanied with approved devices, plenty of researches have further study on polyanhydrides, polyorthoesters, polyphosphazenes, and other biodegradable polymers. Clinical use with the major class of synthetic biodegradable polymers with products was also remained like polyglycolides, polylactides and their copolymers.

An obvious advantage of nature origin based polymers lies in biological macromolecules. The releasing of soluble degradation product could leads to an actual mass loss of the implant where phagocytosis by macrophages and histocytes, intracellular degradation and finally, metabolic elimination through the citric acid (Krebs) cycle to carbon dioxide and water, which are expelled from the body via respiration and urine. The implantation site, the enzyme availability and their action mode comes from the degradation. Natural polymers prevent the chronic inflammation or immunological reactions and toxicity's possibility where implants or drug delivery system was detected with synthetic polymers.

As cellulose, starches, natural rubber and DNA are biodegradable and bioresorbable to support the reconstruction of a new tissue without inflammation are included in the widely variety of polymers.

Fibrinolytic enzymes for immobilization to polymer membranes are chosen from urokinase, streptokinase, fibrinolysine, and brynolase. The enzymes immobilized on membranes have been used clinically.

Some examples for the binary immobilization of antithrombogenic reagents were also provided by some research such as the immobilization of prostaglandin-heparin complex and the urokinase-heparin coimmobilization. According to some articles reported with polytetrafluoroethylene membrane coated with heparin-collagen complex, an early thrombus formation is inhibited by the action of heparin and a longterm endothelialization is accelerated by collagen.

CLINICAL USES

Orthopedics

A quantity number of biopolymers are being used in orthopaedics in substitute of metallic component from long time ago.

Cardiovascular Systems

A great success in heart disease has been proved by biodegradable implants where metallic implant sometimes creates problems. Good results have been shown by biodegradable implants showed better results than metallic ones. Synthetic polymer implants can`t grow and attach with cardiac cells, and in congenital heart diseases they do not give accurate results because of their in-viability. In the replacement of a surgical defect in the right ventricular outflow tract in the heart copolymer of biodegradable poly urethane is generally used. Biodegradable polymers act a good way as a guided tissue regeneration membrane (GTR) etc. in dental problems like filling the cavities. The exclusion of epithelial cells allows the supporting, slower-growing tissue including connective and ligament cells to proliferate.

In sutures and ophthalmic

Suture is a complicated designed and manufactured medical product meet a demand between physical and chemical in a big range which are divided into natural and synthetic wildly.

Biodegradable polymeric nano-particles have been attracted a great interest by many research groups not only in food but pharmaceutical fields as well due to the big usage with favorable properties like good biocompatibility, structure variations, easy design and preparation and interesting bio-mimetic characters. Particularly in the field of smart bioactive carriers, polymer nanoparticles can deliver bioactive compounds into the intended site of action directly. Besides, there are varieties of functional biopolymers and specialized equipment that can be chosen in polymeric nanoparticles producing. In addition, natural nano-carriers like cyclodextrins and caseins have emerged as an attractive option for controlled bioactive systems according to their resemblances with the extracellular matrix in the human body and various other favorable physicochemical properties.

Conclusion

This review detailed definition, classification, types and some important applications of biomaterials used in different applications in our life. From the biomaterials unique properties, these materials are perfect candidates for different bio-related applications.

3

Chapter

BIO-POLYMERS FROM AGRICULTURAL WASTE

INTRODUCTION

This review article explores the production of biopolymers, biodegradable polymers, and polymers from agricultural waste such as fruit seeds, fruit peels, coconut shells, potato peels , orange tree pruning , wheat straw , soy protein isolates , oil palm fiber , sugar palm, corn starch, and rice husks, which are categorized as renewable sources. This review's purpose is to provide conclusive evidence on whether biopolymers, biodegradable polymers, and polymers from agricultural waste were fully biodegradable or only compostable. Current experimental data show that polymers compost at different rates in the environment.

The investigation of green materials such as bio-based plastics is validated by the contribution of synthetic plastics materials to anthropogenic contamination of the environment in each phase of the life cycle from monomer synthesis to disposal in landfills or recycling. The current rate of global plastic production is unsustainable, considering more than 400 million tons of waste is generated each year. Additionally, the rate is expected to increase fourfold by 2050 and there has been a concomitant increase in agricultural plastic waste. The agricultural plastic waste originates from shading nets, mulching materials, and pesticide containers. The volume of agricultural plastic waste would surge in line with the global demand for food cultivated in controlled environments.

The quantity of agricultural waste derived from various supply chains was about 90 million tons of oil equivalent (MTOE). Considering that only a small fraction of the waste is utilized in the production of animal feeds, manure, and other value-added products, there is a potential for the production of biodegradable polymers from agricultural waste. The recycling of plastic waste is not favorable using current technologies due to the risk of leakage of toxic and synthetic chemicals such as anti-oxidants, plasticizers, and stabilizers. The absence of facile, scalable, and environmentally favorable recycling processes has impacted the rate of recycling of global plastics waste—only 9% of the plastics are recycled. The threat of plastics to the environment extends beyond the lack of suitable recycling methods; the synthesis of eco-friendly polymer composites has been impacted by unsuitable synthetic routes.

The rate of non-biodegradable plastic production and land filling coupled with the rapid growth in the global population show that the traditional model, which was primarily based on the extraction of raw materials, production, use, and disposal, is no longer viable in the 21st century and beyond. Environmental advocates have championed the adoption of a new approach to manufacturing that ensures that today's usable products create resources and materials for the development of tomorrow's products; this can be achieved through a circular modern business typology, which integrates agricultural cooperatives, agro-parks, support structures, environmental bio refineries, upcycling entrepreneurship and biogas plants.

Beyond the circular business model typology, sustainability can be enhanced through the production of biodegradable polymers. The current state of research on the production of biodegradable polymers has adopted two approaches. One, biodegradable polymers are manufactured from bio-based precursors, such as agricultural waste, starch, and renewable materials such as poly(lactic acid) (PLA) and polyhydroxyalkanoates (PHA), which are produced by Gram-positive and Gram-negative bacteria. Two, the bio-based polymers are synthesized through the modification of non-biodegradable polymers. The microstructure of non-biodegradable polymers can be modified through the integration of anti-oxidants and pro-oxidant additives, which induce photo-oxidation and oxo-degradation following exposure to ultraviolet light. The utility of the second approach in achieving 100% biodegradation has been contested because non-biodegradable polymers are infused with synthetic stabilizers and photo-initiators, which act as inhibitors in the biodegradation process and UV-oxidation. Considering the limitations of the latter method, the scope of this review is confined to the synthesis of bio-based polymers from renewable sources, especially agricultural wastes. The sources of agricultural waste include post-harvest waste from horticultural plants, sugarcane bagasse, rice husks, and bamboo leave ash.

PRODUCTION OF BIODEGRADABLE POLYMERS

Biodegradable polymers are a unique class of polymers that are ecologically benign (biocompatible and biodegradable). The production process for these biopolymers is grouped into four different classifications depending on the desired products and the available materials/precursors. The classifications are chemical synthesis methods, bacterial synthesis methods, biopolymer blends, and renewable sources. The present discussion primarily focuses on the first type of production, which focuses on the production of bio-based polymers from agricultural waste. Other synthetic routes (biopolymer blends, chemical, and bacterial synthesis) are discussed briefly in the subsequent sections.

The selection of the biodegradable polymers for various commercial applications is based on the physical properties of the polymers. High tensile strength, tensile strength, and yield strength are critical in construction-related applications. In contrast, the % elongation determines utility in packaging. Following the review of the information presented in Table 1, poly(glycolic acid) (PGA) biopolymer has the best tensile strength and modulus of elasticity, but a lower percentage elongation at break. The data show that the mechanical strength is correlated with the density; a higher density translates to better mechanical strength.

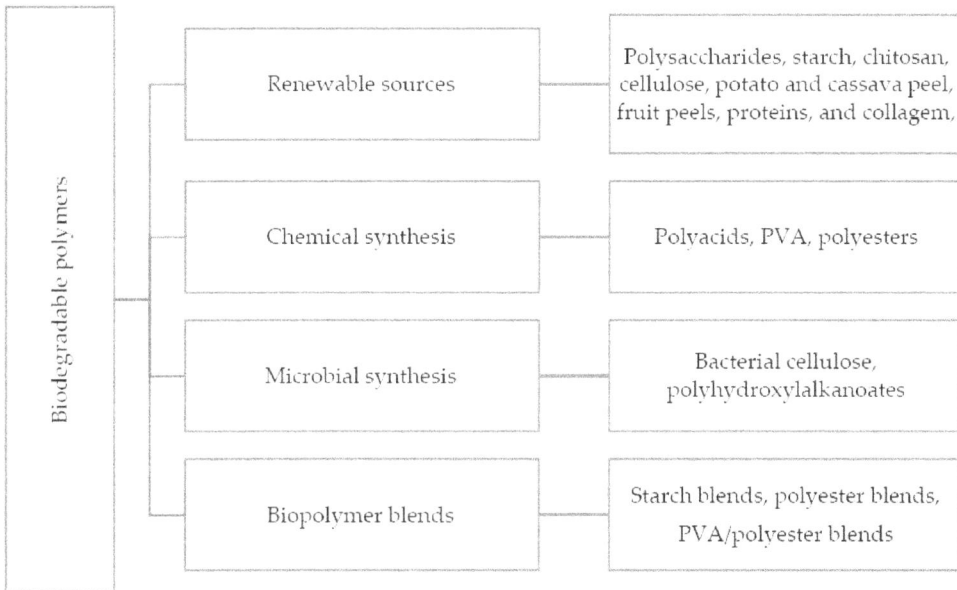

Biodegradable polymers		
	Renewable sources	Polysaccharides, starch, chitosan, cellulose, potato and cassava peel, fruit peels, proteins, and collagem,
	Chemical synthesis	Polyacids, PVA, polyesters
	Microbial synthesis	Bacterial cellulose, polyhydroxylalkanoates
	Biopolymer blends	Starch blends, polyester blends, PVA/polyester blends

Figure 1. Classification of production process for biodegradable polymers

The production of bio-based polymers from agro-wastes is influenced by the availability of starting materials/precursors; these materials should be cheap and available in significant quantities. Among the leading economies, India and China have the capacity to lead in the production of fruit and vegetable-based biopolymers, given the high production capacity and share of total global production. The renewable sources for bio-based polymers are diverse. Bio-based polymers have been synthesized from plant-based precursors containing lignocellulose fibers, cellulose esters, polylactic acid, and polyhydroxyalkanoates (PHA). The lignocellulosic fibers are derived from plants such as curaua, pineapple, sisal, and jute. The physical properties of the final product are largely determined by the extraction method. Organic materials/precursors containing large quantities of cellulose and other fibers are preferred because they enhance the mechanical strength of the materials.

Table 1. Production of Bio-Based Polymers from Renewable Sources & Agro-Wastes

Propertty	Type of Biopolymers						
	PLA	I-PLA	dl-PLA	PGA	PCL	PHB	Starch
Density (kg/m3)	1210	1240	1250	1500	1110	1180	
Tensile strength (MPa)	21	15.5	27.6	60	20.7	40	5.0
Young's Modulues (GPa)	0.35	2.7	1	6	0.21	3.5	0.125
Elongation (%)	2.5	3	2	1.5	300	5	31
Glass transition temperature (°C)	45	55	50	35	–60	5	
Melting temerature (°C)	150	170	am	220	58	168	

On the downside, even though cellulose is a bio-based material, the precursor is non-biodegradable due to the higher degree of substitution. In contrast to other renewables, which are sourced from plants, agricultural waste comprises of post-harvest waste, by-products of food processing such as coconut shells, potato peels fruit peels, and fruit seeds, which have been traditionally discarded as waste in farms and food processing facilities.

Agricultural waste is primary sources of starting materials, which are used in the production of are a vital source of polysaccharides, which are essential precursors in the development of natural plasticizers. The main function of the plasticizers is to enhance the elasticity and mechanical strength of the bio-based polymers. The performance of vegetable-derived polysaccharide plasticizers relative to glycerol and other synthetic plasticizers has not been determined, and commercial application is limited. Agricultural waste such as mango kernel extracts, green tea extracts, essential oils, proto-catechuic acid, grapefruit seed extract, and curcumin sourced from food processing facilities are used in the development of antioxidant additives peel extract (PE), mint plant extracts (ME), Thymus vulgaris L. and oregano. The phenols in the natural antioxidants are Lewis bases and electron donors, which are critical to the anti-oxidation activities. Apart from phenols, pomegranates contain gallic acid and gallates, which are natural stabilizers and indicators of aging.

Bashir noted that these materials had the prerequisite antioxidant activity that was linked to the ability to scavenge for OH groups and oxygen radicals in 2, 2-diphenyl-1-picryl-hydrazyl-hydrate (DPPH). The performance of these materials is comparable to synthetic antioxidants and could, therefore, replace existing additives such as the carcinogenic butylated hydroxytoluene. The main challenge is that the performance of the natural additives on a commercial scale has not been confirmed. The main function of the additives is to inhibit the UV-based photodegradation of the plastics following exposure to sunlight. Apart from the incorporation of natural additives, UV-induced degradation is inhibited by maleic anhydride treatment, direct, reactive mixing, and graft copolymerization during synthesis.

The utilization of waste from renewable sources for commercial purposes has the potential to reduce the rates of global warming, considering that compositing and landfilling contribute to global warming. Data collected from Italy show that the recycling of agricultural waste through composting and the production of fertilizers increases global carbon emissions. In particular, 64 and 67 kg of CO_2 equivalent was generated per mg from olive waste-based compost (OWC) and anaerobic digester-based compost (AD), respectively. Additionally, re-composting and co-composting generated between 8 and 31 kg of CO_2 per mg of compost. The data obtained from the composting experiments show that recycling of agricultural waste poses a threat to the environment, and it is not ecologically beneficial as initially proposed. The significant quantities of CO_2 equivalent emissions generated per mg of compost indicate that novel methods of utilizing agricultural waste such as the production of bio-based polymers are necessary; this because the latter methods are more sustainable and have a lower ecological impact based on the LCA analyses.

Global statistics show that the production of bio-based plastics from renewable sources was low—2.1 million tons were produced in 2018. The projected demand for bio-based polymers would be equivalent to 46% of the global production of packaging plastics by the end of 2020; this translates to put out 7 million tons. The demand for bio-based plastics in food packaging is based on the unique material properties of biofilms relative to synthetic alternatives. The bio-based polymers absorb ethylene, remove water vapor, protect fruits and vegetables from microbial contamination due to the presence of anti-microbial agents, protect against UV radiation, and are easily recyclable.

Current bio-based polymers have shown e_ective antimicrobial performance against Bacillus subtilis, Escherichia coli, and Listeria monocytogenes. The bio-based films have other essential properties that influence the development of intelligent packaging systems.

The ability of current production systems to satisfy this demand is unknown, considering that nearly 50% of the bio-based plastics made from renewable feedstock were non-biodegradable, possibly due to the addition of synthetic plasticizers, and other additives to enhance their mechanical properties. The leading synthetic plasticizers include polyethylene glycol, citrate ester, and oligomeric acid.

Ramesh Kumar, Shaiju, Connor and Babu   noted that global estimates are not entirely accurate due to the complexity of the supply chains, continuous innovation, and commercial release of new varieties of bio-based polymers. The data show that there were two inherent challenges associated with the production of bio-based polymers. Firstly, the production capacity is low, and it cannot match the production of non-renewable plastics, whose production was estimated at 400 million tons.

Secondly, current technologies are limited and inadequate—there are no 100% biodegradable bio-based polymers with optimal mechanical properties..

Considering the global variability in the availability of agricultural waste, the development of the materials would be concentrated in specific geographical areas. For example, fruit peels and coconut shells are found in abundance in tropical and coastal areas, respectively. Since India and China have a high fruit and vegetable production capacity, agro-wastes synthesized from fruit and vegetable wastes would be abundant in Asia. Coconut shells and microalgae are abundant in coastal areas and marine environments, respectively. Jackfruits and other similar plants grow best in tropical and subtropical climates. The data show that the production of bio-based plastics from agro-waste should be customized to suit the available precursors. The development of bio-based polymers from locally available agricultural wastes would also help to reduce the carbon footprint.

Polymers that aremade of poly(butylene adipate-co-terephthalate), poly(butylene succinate/adipate), and poly(e-caprolactone) are biodegradable because the carbon chains are susceptible to enzymatic degradation. Commercially available biopolymers are grouped into the following categories: polylactides (PLA), polyhydroxyalkanoates (PHAs/PHBs), polyols, polyamides, bio-PET, butyl rubber, and cellulose acetate. PHAs are further grouped into long-chain, medium, and short-chain polymers. The length of the chains predicts the utility in commercial applications; short-chain polymers are not ideal in high strength applications owing to their brittleness, high degree of crystallinity, and stiffness. Medium chains are less susceptible to brittle fracturing owing to the high elastic modulus, flexibility (longer elongation at break), and low crystallinity. However, the materials are less suitable for high-temperature applications.

The selection of suitable agro-waste is based on the following primary criteria: (i) starch content; (ii) cellulose and lignin and hemicellulose content (iii) bioavailability and impact on agricultural supply chains and food security (iv) complexity of the synthetic routes and desired material properties; (v) biodegradation . Based on the data presented in Table 2, corn and stalks have the highest cellulose concentration % w/w, which is critical for high strength applications. Experimental data indicate that the production of biopolymers involves a tradeo between the cellulose content and the rate of biodegradation—plant cellulose limits the rate of biodegradation but enhances the mechanical strength of the polymer films—a challenge that has been resolved by Xie, Niu, Yang, Fan, Shi, Ullah, Feng, and Chen. The study reported the successful replacement of plant cellulose with bacterial cellulose. The cellulose and starch content are limiting factors in the selection of agricultural waste precursors.

Table 2. Chemical composition of common forms of agricultural waste

Agro-Industrial Wastes	Chemical Composition (%w/w)					
	Cellulose	Hemicellulose	Lignin	Ash (%)	Total Solids (%)	Moisture (%)
Sugacane bagasse	30.2	56.7	13.4	1.9	91.66	4.8
Rice straw	39.2	23.5	36.1	12.4	98.62	6.58
Corn stalks	61.2	19.3	6.9	10.8	97.78	6.40
Saw dust	61.2	28.1	24.2	1.2	98.54	1.12
Sugar beet waste	26.3	18.5	2.5	4.8	87.5	12.4
Barley straw	33.8	21.9	13.8	11	–	–
Coton stalks	58.5	14.4	21.5	9.98	–	7.45
Oat straw	39.4	27.1	17.5	8	–	–
Sunflower stalks	42.1	29.7	13.4	11.17	–	–
Wheat straw	32.9	24.0	8.9	6.7	95.6	7

Bio-based polymers synthesized from different agro-wastes have distinct material properties.

The starch content in the agro-waste predicts the thickness of the bio-based plastic films—higher starch content is correlated with an optimal thickness (~0.099–0.1599 mm) due to the presence of amylose and amylopectin compounds. Thick films have better mechanical properties compared to thin films.

For example, Chlamydomonas reinhardtii microalgae species yield the highest starch content after 800 h of inoculation. Based on the inoculation experiments, Chlamydomonas reinhardtii microalgae species would be highly preferred as precursors in the development of bio-based polymers compared to other species such as Scenedesmus sp and Chlorella variabilis. Starch content is one of the primary criteria in the selection of the agricultural precursor. The preference for species with high starch content involves a trade with the rate of culture growth. Similarly, higher cellulose content augments the mechanical strength but limits the rate of biodegradation.

Thermoplastic Starch-Based Polymers

Starch is a polysaccharide found in tubers, legumes, and cereals agro-wastes and is an ideal carbon precursor for bio-based polymers. Thermoplastic starch-based polymers are practical alternatives to petroleum polymers based due to effective reinforcement properties, abundance, and tunable properties. The base material, starch (derived from potatoes, cereals, and corn), is abundant in the biosphere and it has been extensively explored in research, as noted by Tabasum, Younas, Ansab, Majeed, Majeed, Noreen, Naeem, and Mahmood. The first phase in the production of starch-based polymers from agro-wastes involves the addition of L-lactate and a catalyst (Sn(oct)2).

Alternatively, the polymerization process can be triggered by the addition of the poly 1, 4-dioxan-2-one (PPDO)–diisocyanate (NCO) group, leading to the formation of starch-g-PPDO polymer chains. Although the process is scalable, the PPDO–NCO + starch/Starch + L-lactide and Sn(Oct)2 reaction results in biopolymers that are easily degraded by water—the addition of plasticizers limits susceptibility to water degradation. The yield (Y%) and grafting effciency (GE) of the starch-PPDO and NCOsynthetic route are determined using the formula depicted in Equations (1) and (2), respectively. W1 denotes the starting weight and final weight. The main challenge with this synthetic route is eco-toxicity. The production of PPDO–NCO relies on 2, 4-Tolylene diisocyanate, and other chemicals that have been proven as toxic to the human body. The use of toxic chemicals impacts the cradle-use-disposal cycle.

Current research has shown that these materials are critical to the future of sustainable food packaging because they are flexible and light. Commercial application is limited by poor water resistance, poor mechanical strength, and risk of dissolution in water—a challenge that is addressed by blending with other polymers to enhance the mechanical strength. Alternatively, TS materials are reinforced by the incorporation of ionic liquids such as 1-butyl-3-methylimidazolium chloride in the pretreatment process and the production of bio-composites. The surface treatment process results in the development of materials with greater activation energies, which predicted the rates of thermal degradation. Biopolymers with cellulose are degraded at temperatures of up to 5000C. The rate of thermal degradation influences end of life treatment and application in high-temperature applications. Other constraints include complex synthetic processes such as plasticizing, casting, and extrusion, which are diffcult to replicate on a commercial scale.

The material property challenges associated with the starch-based polymers are dependent on the starch precursor. Sugar palm, microalgae, and jack fruit result in starch-based polymers with distinct properties, and the synthetic route should be customized to suit the polymer applications. The natural properties of bio-based polymers are modified through the addition of tetraethoxysilane (TEOS), polyvinyl alcohol (PVA), and chitosan. ThePVAis used to enhance mechanical properties a higher PVA ratio compared to the filler was correlated with greater mechanical strength. However, the chemicals (borax and formaldehyde) used in the chemical cross-linking of the biopolymers are toxic and non-biodegradable. Chitosan helps to improve the bonding between natural polymers, TEOS, and PVA. Apart from the material constraints and complex synthetic routes, the sustainability of starch-based polymers is questionable on a commercial scale because starch sources are staple foods in most countries. From a food security perspective, large-scale commercial production of thermoplastics might be a threat to food security. The challenges and viable alternatives in the commercialization of biodegradable polymers are discussed in the next section.

Production of Bio-Based Plastics from Pineapple Peels and Tomato Pomace

The production process of bio-based polymers from pineapple peel is based on a standard method that involves the extraction of biopolymers from agricultural waste. The initial procedures involve the analysis of the chemical composition, especially the C/N and C/P ratios, which predict the polymer yields. Once the numbers of trace metals, ash, and carbohydrates, protein, the peels are fermented (using dipotassium phosphate or ammoniums sulfate) and subsequently hydrolyzed with H2SO4, the biopolymers are extracted via centrifugation at a rate of 4000 rpm or higher. FTIR, NMR, and GC-MS instruments are used in the characterization of the final product. The information presented in Table 2 shows that optimal PHA yields were reported in precursor materials that had the highest content of C/N and C/P. Additionally, the biopolymer yield is influenced by time and pH optimization. The optimal time and pH were 60 h and 9, respectively. The yield data show that chemically induced fermentation was capable of complementing natural bacterial synthesis methods. The only constraint is the possible adverse effect of synthetic chemicals such as H2SO4 and dipotassium phosphate or ammonium sulfate, among other chemicals, which may potentially contribute to acidification and eutrophication in the environment if used in large quantities.

The production of bio-based polymers from tomato pomace follows a similar approach as the production of bio-based polymers from the pineapple peels, except for the melting poly-condensation step. The mechanical properties of biopolymers derived from tomato pomace. Since the linear regression values are close to 1, the linear regression graph confirms that the formation of ester functional groups (COOR-) influenced the hardness and Young's modulus of the biopolymer. The data show that optimal mechanical properties were achieved at 175 0C. The volume of the catalyst (Sn(oct)2) impacted the depth of the indent caused by the Brinnell hardness. Optimal depth was reported in samples with 0.00 mmol of the catalyst.

A contrary phenomenon was noted in the relationship between catalyst (Sn(oct)2), Young's modulus, and hardness. Apart from the production of bio-based polymers, fruit peels are effective in enhancing the mechanical properties of manufactured polymers. Patil, Hrishikesh and Basavaraj observed that the addition of 10–30% of lemon peel powder and sweet lime peel powder reinforces the mechanical strength of natural fibers and epoxy resins. Optimal mechanical performance (Brinnell hardness of 83, the flexural strength of 79 MPa and tensile strength of 48 Mpa) was reported in the epoxy-lemon biopolymer, which had a 30% volume weight of lemon particles. The improvement in the mechanical properties was linked to the presence of cellulose, lignin, and crude fibers, which made up of 90% of the sweet lime and lemon. Additionally, there was good particle distribution and particle-matrix adhesion. Even though the sweet lime and lemon peel showed ideal properties in the reinforcement of the structures, the sustainability aspect remains a challenge; this is because the lemon and sweet lime fruits are edible and the commercial availability of waste fruit peels is not guaranteed. In advanced markets, the fruit peels are used to produce value-added products such as bioactive polyphenols.

Phenol-containing compounds have natural antioxidant capabilities. Alternatively, the peels are ingredients in the manufacturing of home-based beauty products. The production of lactic acid and poly-lactic acid from agro-wastes is discussed in the next section.

Production of Lactic Acid, PLA and PHA from Agro-Wastes

The production of lactic acid and poly-lactic acid is influenced by specific strains of bacteria for fermentation and hydrolysis and the availability of agro-wastes as starting materials. The fungi and bacteria strains adopted for commercial applications include Rhizopus, Pediococcus, and Streptococcus.

The availability of a wide array of bacteria and fungi species has an impact on the material properties (biochemical characteristics, morphological, and psychological characteristics) of the final product due to the variations in the fermentation processes that lead to the production of fermentable sugars such as starch and cellulose. Apart from the utilization of different strains of bacteria, the material properties of the PLA- and lactic acid-based polymers are influenced by the pre-treatment methods (cold and thermal) that are primarily used to remove undesired materials. The fermentation process results in the formation of lactic acid, which is polymerized to form PLA. Biopolymers that are synthesized from agricultural wastes have a tensile strength of 36.3 MPa and a melting point of 170OC. The high tensile strength and melting point show that the polymers are suitable for packaging applications and agricultural shading.

Production of Bio-Composites from Winery Agro-Wastes and Sugar Beet

Merlot grape pomace fruit waste is the main form of winery agro-waste. In place of decomposition, the winery agro-wastes are a suitable source of composites that are manufactured through solvent extraction (SE) methods, and pressurized liquid extraction (PLE). The extracts drawn from PLE and SE methods are mixed with commercial-grade polyhydroxyalkanoate to form the matrix. The final phase of the production involves mixing the biopolymer with the poly(3-hydroxybutyrateco-3-hydroxy valerate) (PHBV)—a copolyester containing hydroxyaleric acid to form active bio-composites.

The bio-composites have higher or higher than normal tensile strength compared to the virgin biopolymers or the matrix in isolation. The highest mechanical strength was reported in the virgin PHBV matrix. The inclusion of the bio-based materials extracted via solvent extraction resulted in a reduction in the tensile strength and a marginal improvement in the elongation at break. The data also show that the synthetic route/extraction method for phenols had an impact on the mechanical properties of the bio-composites—solvent extraction (SE) was a practical solution compared to pressurized liquid extraction (PLE).

Beyond grape pomace, sugar beet agro-wastes are practical sources of bio-composites owing to the presence of carbocal in the dried pulp. The mechanical properties of the Carbocal are enhanced through the formation of an LLDPE-carbocal biopolymer, via mixing, sieving, drying, and injection mounding. An analysis of the mechanical properties showed that higher carbocal content improved Young's modulus but compromised the elongation at break. There were limited necking and plastic deformation.

Chemical and Microbial Synthesis and Chemical Extraction

Biodegradable polymers are also generated by the activity of microorganisms such as Gram-negative and Gram-positive bacteria in the presence of carbon-rich materials such as agro-wastes. The bacterial production of the polymers is triggered by pH changes, limited availability of essential nutrients such as phosphorous and nitrogen, the composition and type of culture, and media. The naturally occurring biopolymers act as biological storage systems or defense mechanisms. Microalgae are critical to the biological storage processes that result in the development of biopolymers, through biological carbon fixation via photosynthesis. The process culminates in the formation of branched polysaccharides. PHA is the leading bio-based biopolymer that is synthesized from microbes. The synthesis of bio-based polymers from rice bran is catalyzed by the microbial activity of Sinorhizobium meliloti MTCC 100 bacteria. These bacteria are preferred compared to other species and synthetic methods because they do not pose a threat to the environment and generate significant quantities of agro-wastes. The microbial synthesis method generated PHA, biomass, and exo-polysaccharides (EPS) at a rate of 3.63, 1.75, and 1.2 g/L, respectively. The rate of production was augmented by the optimization of the incubation period and the addition of rice bran hydrolysate

(RBH) at predefined intervals in the fermentation process. Fermentation time, temperature, and pH optimization experiments showed that optimal conditions for the synthesis of PHA biopolymer were neutral pH conditions, 30 OC, and 72 h. Even though the mechanical properties of the polymer were not measured, the FTIR spectra confirmed the presence of C=O, CH, and C-O-C; these functional groups are associated with hydrogen bonding and advanced chemical bonding that help to predict the mechanical strength and the presence of specific functional groups such as cellulose and lignin.

Other microbes, such as white rot fungi, help in the natural de-lignification of agro-wastes. Microbial synthesis methods have also proven effective in the production of poly b-hydroxybutyric acid (PHB)—a biodegradable and high strength PHA biopolymer. The bacterial synthesis of the biopolymer is dependent on the availability of a carbon-rich precursor that is utilized by the bacteria as a source of food and energy. In contrast to other microbial synthesized biopolymers, PHB is suitable for high strength applications because it has mechanical properties that are nearly identical to petroleum-based biopolymers such as PP. The primary constraint is the cost, which is nine-fold higher compared to other biopolymers. The cost is attributed to the market price of the carbon-rich starting materials.

Table 3. Mechanical properties and optical properties of microbial synthesized starch films

Type of Material	Optical Trans mission	Solution (%)	Tensile Stress at Break (MPa)	Tensile Strain at Break (mm/mm)	Thickness (μm)	WVP (g.mm/ Kpa. m² h¹)
Control crystalline nanocellulose	24.0 ± 3.10	15.19 ± 0.11	3.1 ± 0.39	0.35 ± 0.08	199 ± 26	1.9 ± 0.03
Bacteriocin (from P. acidilactici	64.4 ± 204	20.73 ± 0.05	3.3 ± 0.45	0.35 ± 0.04	183 ± 27	1.78 ± 0.06
Bacteriocin (from E. faecium)	63.2 ± 215	11.60 ± 0.20	2.85 ± 0.52	0.44 ± 0.02	195 ± 29	1.70 ± 0.06
BIN (bacteriocin from P. acidilactici) BIN	62.2 ± 4.78	12.32 ± 0.21	3.04 ± 0.50	0.44 ± 0.05	198 ± 23	1.69 ± 0.03
(bacteriocin from	53.9 ± 2.74	21.54 ± 0.51	4.33 ± 0.29	0.29 ± 0.03	187 ± 36	187 ± 0.04
E. faecium)	52.1 ± 258	22.2 ± 0.18	5.21 ± 0.53	0.30 ± 0.02	187 ± 20	1.72 ± 0.04

The cost-related factors have been resolved through the utilization of agro-wastes, such as wastes drawn from rice and jowar processing. The utility of different strains of fungi and bacteria in agro-waste biopolymer synthesis shows that the quality of biopolymer produced was dependent on the types of cultures used and the media.

WVP denotes—Water vapor permeability; BIN—Bacteriocins immobilized crystalline nanocellulose (BIN).

Unadulterated cultures were associated with higher volumetric productivity and high costs. In contrast, mixed cultures were affordable but resulted in poor yields. Apart from the yield, the type of microbes predicted the rate of biodegradation, the rate of biodegradation oscillated between 46% and 63%. The maximum rate of biodegradation was reported in bio-based polymers made from crystalline nanocellulose derived from agricultural sources. The presence of *E. faecium* resulted in the most significant reduction in the biodegradation rate but slightly higher tensile strength at break compared to the *P. acidilactici* species. The microbes also impacted the optical properties and water vapor permeability (WVP). Biopolymers with minimal optical transmission % were ideal for greenhouse-related applications. Advanced methods involving genome sequencing have facilitated the synthesis of customized bacterial biopolymers such as PHB and PHA from recombinant *E. coli* and other microbes. Advanced genetic methods have also facilitated the customization of the plant composition—the natural variability in plant cuticle distribution (raw materials for bio-based polymers) has been resolved by advanced breeding methods.

Chemical synthetic methods involve the treatment of agro-wastes/food wastes with acids and alkalis to extract the lignin and cellulose materials. The treatment processes form functional groups (C=O, and C-O-C, among others), which influence the mechanical properties of the biopolymer through the strength of the chemical bonds. On the downside, the chemical process generates furan derivatives, carboxylic acids, and lignin-derived phenols that inhibit the enzymatic activity, which is critical in the fermentation, phase the negative impact of acids and alkalis on the fermentation process can be reversed by the incorporation of specific species of white-rot fungi such as Ceriporiopsis subvermispora in the pretreatment process. The use of microbes in chemical synthesis does not negate the fact that synthetic chemicals are harmful to the environment and diminish the essence of using agricultural precursors in place of PE, PET, and PP.

CHALLENGES IN THE PRODUCTION OF BIOPOLYMERS POLYMERS AND AVAILABILITY OF PRECURSORS

Kumar and Kumar noted that available routes are characterized by an incompatibility between the hydrophilic water fibers and the hydrophobic polymer matrix, which is water repellent. The lack of compatibility leads to uneven dispersion and low mechanical strength. The inability to match the mechanical properties of non-biodegradable polymers limits the utilization of biodegradable polymers to applications that require low mechanical strength. The enhancement of the mechanical strength involves a tradeo_ with biodegradability—the biodegradable materials have to be blended with polymers to enhance their mechanical strength. Additionally, biological precursors such as cellulose acetate that yield high tensile strength (90 MPa) are not biodegradable. The limited rate of biodegradation has been addressed in recent studies by replacing plant cellulose with bacterial cellulose, which has ideal water holding capacity and better biodegradation rates. The unique properties of bacterial cellulose are associated with the ultrafine nano-fibrils in the 3D network structure. The variations in the mechanical properties of bio-based polymers and petrochemical-based polymers are presented in Table 4. The data show that biodegradable materials such as PHA have limited tensile strength, elongation at break, and glass transition temperature compared to PET and PE.

Table 4. Comparative analysis of the mechanical properties of plant-based and petro-chemical based polymers .

Material	tensile Strength (Mpa)	Elongation at Break (%)	Glass Transition Temperature (°C)	Melting Temperature (°C)
Kraft paper	68	3		
Cellulose acetate	90	25	110	230
Corn starch	40	9	112	
PLA	59	2-7	55	
PHA	15-15	1-800	12-3	100-175
PBS	34	560	32	114
PBAT	22	800	29	110
PEE	35-67	3-4	85	211
PTT	49	160	50	228
PE	15-30	1000	125	110-130
PP	36	400	13	176
PET	86	20	72	265
PS	30-60	1-5	100	-
PVC	52	35	18	200

Apart from the material-related shortcomings, the production of plastic materials is not economically viable compared to standard plastics. From an economic dimension, biodegradable plastics are not sustainable, considering that cost is a critical criterion in commercial applications. The data show that plant-based polymers such as PHA are four times more expensive relative to conventional polymers. The bio-based polymers cannot compete with standard plastics in the commercial market because consumers make purchase decisions based on the value of a product relative to the price. The cost factor can be attributed to the economies of scale and limitations in viable technologies. Biodegradable polymers are produced at a smaller scale—99% of plastics (equivalent to about 335 million tons) produced for commercial applications are either non-biodegradable or partially biodegradable. The lack of scalable technologies has also influenced the pricing of these materials.

PRODUCTION AND MARKET SUSTAINABILITY OF BIO-BASED POLYMERS

The sustainability of bio-based plastics is dependent on an array of factors and criteria. The most critical are (i) the availability of commercially viable quantities of renewable feedstock and agricultural waste; (ii) scalable and facile production routes; (iii) cost and competition with synthetic polymers; and (iv) useful life and biodegradation/end of life treatment. Sustainable bio-based polymers should satisfy each of the four criteria. Even though empirical evidence suggests that agricultural waste is available in significant quantities, there has been an inadequate assessment of the availability of agricultural wastes that can help meet the global demand for bio-based polymers, especially in food packaging.

AVAILABILITY OF COMMERCIALLY VIABLE QUANTITIES OF RENEWABLE FEEDSTOCK AND AGRICULTURAL WASTE

FAO global estimates of the quantities of cereals, oilseeds, and pulses, roots and tubers, fruits, and vegetables that are lost each year globally suggest that 20–30% of fruits and vegetables are lost in farming (agriculture) and post-harvest phases across all continents. India and China have one of the highest rates of fruits and vegetable wastes, estimated at USD 484 million per year. The financial losses are linked to the loss of 30–40% of fruit and vegetable produce. Cumulatively, India, UK, China, Mexico City, and Central de Abasto generate about 60 million tons of fruit and vegetable wastes each year. Additional losses occur at the point of sale due to handling and poor consumption habits. Industrialized Asia had the lowest level of food waste. The estimates are slightly lower compared to those reported by Alexander, Brown, Arneth, Finnigan, Moran, and Rounsevell. The research estimated that wastes, losses, and ine_ciencies in the supply system accounted for 44% of global food. The main question is whether the significant quantities of food that were discarded as waste are available in centralized locations. The availability of agricultural waste in centralized locations is critical, given the fast rates of biodegradation.

The FAO estimates on pulses and seeds that are lost in various agricultural supply chains and the level of wastage are lowest in industrialized Asia, North America and Oceania, and Europe. The FAO estimates that 1.3 billion tons of food is wasted annually during farming and post-harvesting and agricultural processing. An EU-28 survey conducted between 2010 and 2016 estimated that 118 billion tons of agricultural wastes, co-products, and by-products (AWCB) were generated during that period. Considering that 68% of this waste originated from fruits, cereals, and vegetables, the waste was a potential starting material for the production of bio-based polymers.

On the downside, the wastage occurs at the consumer or processor level, which limits the possibility that the agricultural waste would be collected and channeled towards the production of bio-based polymers. In general, the volume of waste does not predict the availability of agricultural waste for conversion into biopolymers because there are other competing applications such as composting, bio-fertilizer, and biogas production. The Waste and Resources Action Programme (WRAP) and other social organizations across the EU are advocating for the responsible use of food to reduce the volume of food wastes—so far, these efforts have resulted in a 15% reduction in waste. If these efforts are sustained, food/agricultural waste will reduce significantly.

SCALABLE AND FACILE PRODUCTION ROUTES

Current methods used in the production of bio-based polymers are not adequately scalable. The production of PHA from fruit peel relies on west chemistry methods, whose level of efficiency is dependent on the physicochemical parameters; the yield is not consistent. The production of bio-based plastics from selected sources requires extensive and advanced processing, which consequently impacts the cost of the material. For example, coffee grounds are highly hydrophilic and chemically incompatible with hydrophobic copolymers. The compatibility between these materials is augmented by the addition of coupling agents and compatibilizers. Alternatively, the materials are subjected to thermal treatment under a vacuum environment to improve the hydrophobic properties. The limitations of coffee grounds show that not all agro-wastes are ideal precursors for the development of bio-based plastics.

The production of bio-based polymers from starch and biopolymer blends relies on techniques that have limited commercial utility. Mater-Bi/Novamont, Minerv-PHATM, and Bio-Onhave, among other companies, adopted these methods in the production of bio-based polymers. However, the production capacity is unsatisfactory (97–560 kilotons). Another constraint is that optimal performance has been reported in biopolymer blends made of bioethanol, among other products that are not 100% biodegradable. The production of bioethanol competes with the human food supply chain and might increase the possibility of food insecurity; this is a common challenge for biofuels. Emerging reports suggest that the production methods would be augmented by the development of new synthetic/production routes. These promising methods are based on either pilot studies or applications that have not been proven on a commercial scale, such as the BBI-EU partnership. Based on published evidence, there are limited facile and scalable methods for the production of bioplastics.

Diverse biopolymers can be developed from more than 100 types of available agro-waste. Additionally, diverse composites can be developed by integrating carbon fibers, cellulose, and natural anti-oxidants. Tabasum, Younas, Ansab, Majeed, Majeed, Noreen, Naeem, and Mahmood documented 88 biopolymer types that can be developed from corn starch and natural, synthetic polymers. Each type of material had different characterization techniques and applications.

From a production perspective, the diversity of the biopolymer materials that can be developed from available agro-wastes has a mixed impact. On the one hand, the mechanical, optical, and chemical properties can be customized through the addition of nanoparticles, and copolymers. On the other hand, the production of diverse biopolymer materials limits commercial applications because the imperfections in the new biopolymers cannot be resolved simultaneously. The limitations of modern synthetic routes could be the main reason why the production of bio-based polymers is unable to match synthetic polymers. The production of bio-based polymers was 2.1 million tons.

APPLICATIONS OF AGRICULTURAL WASTE-DERIVED BIOPOLYMERS

The main applications of interest are in food packaging, construction, and agriculture. The applications are influenced by the mechanical, physical, and chemical properties of the material. High strength applications in agriculture and construction require biopolymers with significant tensile strength/Young modulus. In contrast, flexibility/elongation at break is a key criterion in packaging applications. A compilation of different mechanical, physical, and chemical properties of commonly available agro-wastes showed that tamarind fruit fiber has the best mechanical properties (tensile strength of 1137–1360 MPa), making it an ideal source of biopolymers for construction applications and a viable alternative to synthetic carbon. The data show that there is a relationship between the mechanical, physical, and chemical properties.

APPLICATION IN CONSTRUCTION AND AUTOMOBILES

The application of bio-based polymers in construction is influenced by the reinforcing materials such as carbon nanotubes (CNTs), carbon nanofibers (CNFs), nanocellulose, cellulose, lignin, hemicellulose, and α-cellulosic micro filler derived from agro-wastes. The reinforcement of bio-based polymers is critical because the materials have high water permeability rates and are biodegradable. Advances in material science and nanotechnology have facilitated new applications in the construction sector.

The reinforcement of bio-based polymers with carbon nanofibers (CNFs) and carbon nanotubes (CNTs) has increased the suitability of the rice-husk derived polymers in construction applications. The changes in the surface and cross-sectional morphologies before and after coating with CNFs are illustrated in Figures 3 and 4. CNTs are preferred because they have high tensile strength (7 GPa) and Young's modulus (400 GPa) compared to biopolymers alone. The tensile strength of bio-based polymers such as PLA and PHA is below 100 MPa. Additionally, the CNTs and CNFs have a wide aspect ratio and can be easily dispersed into the biopolymer-cement mixture.

The changes in surface and cross-sectional morphology after applying a coat of CNFs on rice husk ash. The presence of nano-scale fibers on the surface of the risk husk ash contributed to greater mechanical performance. In particular, a 15% modification of the risk husk ash resulted in a 187% improvement in the compressive strength and flexural strength after 28 days. On the downside, the incorporation of the CNTs has undesirable effects on the concrete matrix—the higher van der Waal forces and specific surface energy in the composite lead to the agglomeration of the CNTs, a phenomenon that causes greater bridging effects and crack growth within the composite.

The mechanical benefits afforded by the development of biopolymer-cement-CNT composite outweigh the risk of agglomeration and crack growth because agglomeration is reversible through ultrasonic and high-speed shear dispersion. Apart from CNFs and CNTs, the mechanical properties of rice husks can be modified by copolymer blending with PE and treatment with diazonium salt

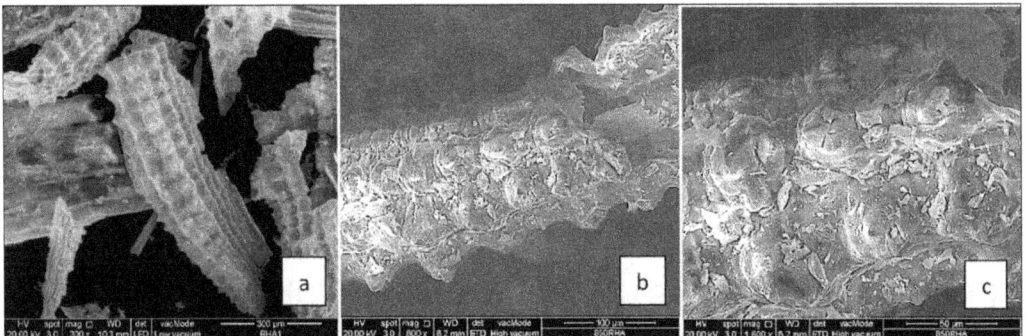

Figure 3. Surface and cross-sectional morphology of rice husk ash biopolymer before coating with CNFs, (a–c) for different scales

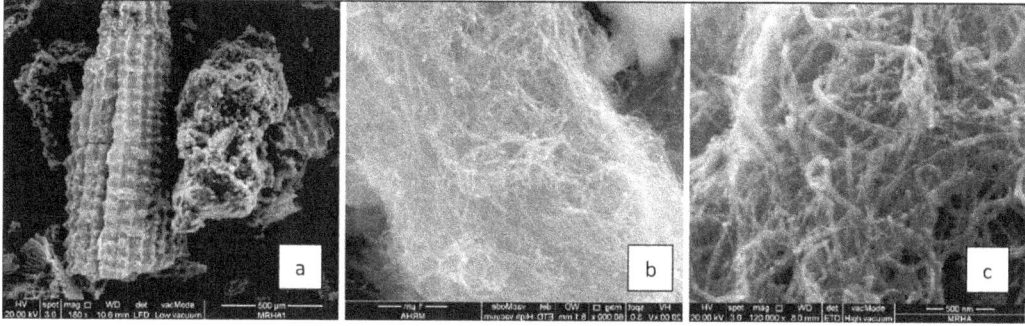

Figure 4. Surface and cross-sectional morphology of rice husk ash biopolymer after coating with CNFs, (a–c) for different scales

Polymer Matrix Composites (PMC) from Agro-Wastes

The application of biopolymers in construction applications is supported by the formation of polymer matrix composites from agro-wastes. The main sources of the agro wastes are grape stalks, olive pits, and wet olive husks, sweet lime, and lemon peels. The lime and lemon peels had an optimal tensile strength of 48 MPa. The mechanical strength of the PMCs derived from grape stalks, olive pits, and wet olive husks was attributed to the higher composition of cellulose, lignin and hemicellulose and the presence of basic and acidic groups, such as carboxyls, lactones and phenols, that are bonded through inter-and intra-molecular hydrogen bridge links. On the downside, the reinforcement of the mechanical properties compromises the end of life treatment of the materials. PMCs and other composites that contain lignocellulose materials have higher thermal stability. The graph shows that the lignocellulose sample had a thermal degradation range of 362–694OC. The improvement in thermal behavior is related to high carbon ratios. Even though the lignocellulose materials are associated with high tensile strength, the utilization of the materials has significant environmental drawbacks, including incompatibility with commercially available biopolymers, poor wettability, and high rates of humidity absorption, which may compromise the integrity of the concrete structures.

Composites made of α-cellulosic micro fillers and epoxy matrixes have been used in construction applications to replace wood, and the replacement of interior metallic door panels in BMW and Mercedes-Benz branded automobiles. The α-cellulosic micro fillers are synthesized from agro-waste materials such as date seeds, Robusta coffee, coconut shells, wood, oil palm shells, walnut, hazelnut, and red coconut empty fruit bunch. A comparative analysis of the performance of different materials shows that the coconut empty fruit bunch has comparable tensile strength as commercial cellulose—52 MPa. The tensile strength is significantly higher relative to lignocellulosic and short fiber fillers made of oil palm shells, nuts, and banana, respectively. Additionally, the impact strength was higher compared to cellulose (0.85 versus 12.8 kJ/m3). The mechanical properties show that α-cellulosic micro fillers are suitable in high strength applications.

IN AGRICULTURAL SHADE NETS (ANTI-INSECT NETS) AND MULCHING FILMS

Bio-based polymers made of cellulose, starch; polyhydroxyalkanoates (PHA), bio-polyethylene, and PLA are employed in the manufacturing of agricultural shade nets and mulching films. The shade nets are vital in integrated pest management due to the toxicity of commercial pesticides—reducing the use of pesticides has ecological and economic benefits and better mechanical properties compared to the traditional LDPE films. Additionally, the nets help to filter UV radiation, which is harmful to plant growth. The commercial application of these nets is influenced by tensile strength; mesh sizes surface color, and chemical composition. Shade nets with high tensile strength have a longer useful life and are capable of withstanding meteorological hazards such as strong winds, sunlight, and hail.

APPLICATION IN FOOD PACKAGING

Bio-based biopolymers that are effective in food packaging applications are PLA, sugar palm nano-fibrillated cellulose (SPNFCs), coffee grounds-PBAT composites, blueberry agro-industrial waste, and corn starch. The choice of different agro-wastes in the production of food-packaging materials is based on the sustainability considerations. The poor mechanical properties of PLA do not impact food packaging applications where tensile strength is not a critical factor. The low carbon footprint of PLA and other beneficial ecological effects show that PLA has the potential to replace polypropylene and polystyrene, among other non-biodegradable plastics used in packaging. In general, the poor mechanical strength of unblended bio-based polymers coupled with the high rates of biodegradation and water permeability does not impact packaging applications in the food sector. According to Soares, Siqueira and Prabhakaram, bio-based polymers made through electrospinning/electrospray technology have found new applications in food packaging, tissue engineering, drug delivery systems, wound dressing, and enzyme immobilization.

Coconut-fiber-based biopolymers have been used to develop handicrafts and gardening products. The utility of agro-waste based packaging films has been enhanced by surface modification using nano-fibrillated cellulose concentrations. Ilyas, Sapuana, Ibrahim, Abral noted that the modification of agro-waste biofilms made of sugar palm with nano-fibrillated cellulose concentrations resulted in a significant improvement in the physical, mechanical, and morphological properties. The mechanical properties of sugar palm nano-fibrillated cellulose (SPNFCs) was influenced by the cellulose content—an increase in the cellulose content from 0.1 to 1.0 wt% translated to a greater improvement in the mechanical strength. The optimal ratio of SPNFCs was one—for modulus of elasticity and tensile strength. However, virgin sugar palm starch had the best performance in terms of elongation at break. The data show that higher concentrations of cellulose reduce the flexibility of the biopolymer.

The improvement in the mechanical characteristics can be linked to the microscopic changes that occur during the formation of the composite. The FESEM and TEM micrographs showed significant pore deformation, poor crack formation, and reinforcement of the matrix following the application of the starch coating. The FESEM and TEM data were augmented by the FTIR data, which confirmed the presence of C=O, C-O-C, and O-H functional groups based on the peaks that were observed at 995, 1335, and 1644 cm-1. In other studies, the presence of carboxyl acid groups, phenols, and lactones was associated with the formation of inter and intra-molecular hydrogen bridge links. The presence of these functional groups confirmed that there was extensive hydrogen bonding, which translated to higher chemical bonding between the starch molecules and the SPNFCs. The XRD diffraction patterns showed that there was a considerable improvement in the relative crystallinity that was directly linked to the addition of the starch. In brief, the mechanical strength was linked to changes in the chemical composition that occurred during the development of the composite.

Modified coffee grounds have been used in the synthesis of bio-based films for packaging because virgin materials have limited hydrophobic properties that limit blending with synthetic polymers. The material constraints of coffee grounds are resolved through the use of alternative reinforcing materials such as organo clay-based bio-nanocomposites, chitosan, carboxyl methylcellulose, polylactic acid, and lignocellulosic reinforced materials. There was also the formation of a PMC material from rosin/expanded rosin organoclay (ROC) and PLA and PBAT. The PMC synthetic route is an integral route to the development of commercial-grade biopolymers from agro-wastes. Coffee grounds are poor sources of biopolymers due to the hydrophobicity properties. The hydrophilic nature of cofee grounds impacts compatibility with hydrophobic polymers, limiting the bio-refinement related applications.

The challenge has been resolved through the development of the polymer matrix, chemical or microbial treatment with coupling agents, and compatibilizers. The compatibilizers can be replaced by bio-reinforcing agents, which improved the tensile strength and mechanical performance of the material relative to untreated materials. The stress–strain curve indicates that the treatment of virgin PBAT with a coffee grounds ratio of 10 yielded the best tensile strength performance at both 250 and 270 OC. The modification of the biopolymer structure of blueberry powder and corn starch biopolymers via photo bleaching contributed to the intelligent food packaging systems. The luminance values (surface color) of the biofilms were diminished by photo-bleaching and were ideal colorimetric indicators for packed food deterioration. The biofilms turned blue and red in acid and basic pH environments, respectively. The deterioration of packed food is characterized by fermentation and an increase in the pH. The color changes can be discerned by the human eye. On the downside, the intelligent pH detection data are inconclusive because they are not correlated with the shelf life of foods containing different biomolecules such as proteins, lipids, salt, and sugar.

USEFUL LIFE AND BIODEGRADATION/END OF LIFE (EOL) TREATMENT

The length of the useful life of bio-based polymers is dependent on the level of exposure to UV radiation, which induces photo-oxidation, heat-induced thermal degradation, risk of dissolution in water and mechanical strength in high strength applications. There are diverse options for the end of life treatment of biodegradable polymers; these include home decomposition, industrial composting, enzymatic depolymerization, catalytic recycling, chemical recycling, mechanical recycling, and anaerobic digestion. The choice of each method of EoL treatment is informed by the type of precursor. For example, PLA is primarily recycled via mechanical recycling, chemical recycling, or industrial composting. The options available have helped to mitigate the risk of global warming and carbon emissions. However, these methods are not 100% recyclable. A biodegradation rate of between 60% and 80% has been reported in previous studies. Cellulose-based biopolymers achieved a 70% biodegradation rate after 350 days, while a similar rate of biodegradation was reported after five months in agro-based composite materials. The rate of biodegradation is also contingent on the microbes used in microbial synthesis. Biopolymers synthesized using *P. acidilactici*, and *E. faecium* had the lowest biodegradation rates (<50%). The addition of reinforcing agents resulted in a further decline in the rate of biodegradation. The limited rates of biodegradation raise fundamental questions on whether the materials classified as "biodegradables" are truly biodegradable or only compostable. The ISO 14,855 standards indicates that a material satisfies the biodegradability criteria if 90% of the initial mass is lost within 6 months at 59OC. Additional provisions under ASTM D5338 indicate that biopolymer blends are biodegradable if they achieve 90% loss in mass within 180 days. The acceptable rate of mass loss of homopolymers after 180 days is 60%. A material that does not satisfy the biodegradable criteria can be categorized as compostable because compostable plastics are biodegradable but biodegradable plastics might not be compostable; mass losses define the distinction between biodegradation and compostability.

The main concern is that the rate of biodegradation varies widely depending on the environment (marine, soil and freshwater). Other concerns include limited data on biodegradable polymers and polymers derived from agricultural wastes that satisfied these criteria. Despite the paucity of information, it is evident that cellulose-based biopolymers and materials made by *P. acidilactici*, and *E. faecium* did not meet the ISO criteria for biodegradation. The risk to the environment is not eliminated even in 100% biodegradable polymers, because nano-scale materials originating from the degraded polymers have the potential to trigger water and air pollution.

Compostability of PLA and PHA Biopolymers

Experimental data do not provide conclusive evidence on whether PLA and PHA biopolymers are fully biodegradable or only compostable. Based on the ISO definition, both PLA and PHA polymers are not 100% biodegradable. Recent experiments noted that PHA/PLA materials achieved a 68–72% mineralization in 90 days; there is no further evidence on whether the rates improved in the post-90 day period or whether the rate of degradation stagnated. Even though PHA/PLA biopolymers do not satisfy the biodegradability criteria, Emadian, Onay and Demirel classified these materials as biodegradable. In brief, there is no consensus among researchers on compostability and biodegradability of biopolymers, biodegradable polymers, and polymers from renewable agricultural waste.

CONCLUSION

This chapter yielded new knowledge on the production of biopolymers, biodegradable polymers, and polymers from renewable agricultural waste sources such as grape and tomato pomace, green tea extracts, essential oils, and curcumin, coconut shells, vegetable waste, rice husks, fruit peels, grapefruit seed extract, waste vegetables, maize and wheat starch, and municipal agro wastes.

Sustainability is a primary criterion that influences the choice of the precursor (the type of agro-wastes). The production of biopolymers requires commercially viable quantities of agro-waste—a key challenge considering that the wastes occur at the retail and household levels, and there is no mechanism for sorting and disposal of the wastes. Additionally, there is a global variability in the availability of agro-wastes, a factor that influenced the mechanical and optical properties of the polymers developed.

Fruit peels and coconut shells are common in fruit growing regions in tropical and subtropical areas, and coastal areas, respectively. Grape pomace waste is available in regions with grapevines such as Italy. The variations in the availability of waste impact the rate of production. Another constraint is the lack of facile and scalable synthetic routes. New and novel methods are based on laboratory models or experiments, which have not been applied on a commercial scale. Commercial methods include copolymer blending and chemical synthesis; these methods led to the formation of biofilms and bio-plastics, which are not 100% biodegradable. The reinforcement of the mechanical properties involves a trade-off with the elongation at break, thermal degradation ability (at the end of life treatment), and ecological impact, including carbon footprint, and eco-toxicity.

The adverse impact of chemical additives, stabilizers, and photo-initiators has been ameliorated through the development of bio-based anti-oxidant additives made from agro-wastes such as mango kernel extracts, green tea extracts, essential oils, proto-catechuic acid, grapefruit seed extract. The limited synthetic methods available have impacted the costs and the ability of the bio-based plastics to favorably compete with synthetic polymers in the market. The cost factor partly explains why the global market share of bio-based plastics is below 1%. Other emerging concerns include the end of life treatment and useful life—natural bio-based polymers are susceptible to water attack and lack appropriate mechanical strength. Surface doping, blending with commercial polymers and the formation of polymer matrix composites improve the mechanical strength and reduce the rate of biodegradation. The current state of research and development in the production of bio-based plastics predicts the future of bio-based plastics and contribution to global sustainability. The progress made in the production of bio-based films through electrospinning/electrospray technology, nano fibrillated cellulose concentrations, and reinforcement with cellulose has contributed to the demand for biofilms in packaging. The utility of biopolymers in construction and agricultural applications is contingent on the availability of synthetic methods that balance between the tensile and flexural strength, biodegradation, and ecological impact.

4

Chapter

BIO-BASED THERMOSETTING RESINS COMPOSED OF CARDANOL NOVOLAC AND BISMALEIMIDE

INTRODUCTION

In recent years, renewable resources-derived polymers (bio-based polymers) are attracting a great deal of attention because of the advantages of these polymers, such as conservation of limited petroleum resources, possible biodegradability and the control of carbon dioxide emissions that lead to global warming. Most recently, much focus is being placed on bio-based thermosetting resins, such as epoxy resins, phenol resins, unsaturated polyester resins and their composites, because these materials are hard to be recycled owing to the infusible and insoluble properties.

Petroleum-based thermosetting bismaleimide resins are used as the matrix resins for multilayer printed circuit boards and advanced composite materials in the aerospace industry. For example, a commercial thermosetting bismaleimide resin, Matrimid 5292 (Ciba Geigy, Basel, Switzerland), based on 2,20-diallybisphenol A (DABA)/ 4,40-bismaleimidediphenylmethane (BMI) is one of the leading matrix resins in carbon fiber composites for advanced aerospace application. The Fourier transform infrared spectroscopy (FTIR) analysis of the cured DABA/BMI resin has shown that a stepwise end reaction and subsequent chain polymerization in addition to an etherification reaction occur for DABA/BMI. We had already reported the thermal and mechanical properties of the BMI cured with drying oil such as dehydrated castor oil and tung oil.

Cardanol (CD) is a promising bio-based feedstock, which is obtained from anacardic acid, the main component of cashew nutshell liquid, a by-product of cashew nut processing. CD is a phenol meta-substituted with a long unsaturated hydrocarbon chain (C15H31-2x: x=degree of unsaturation; 0–3, average x=2.1), which is expected to undergo the end reaction and subsequent Diels–Alder reaction with BMI (Figure 1). There have been many reports in the literature on the utilization of CD to phenol resins and epoxy resins. However, there are only a few reports on the application of CD by use of the reactivity of the unsaturated hydrocarbon chain. To the best of our knowledge, the curing reaction of CD and maleimide has not yet been reported. Herein, the thermal and mechanical properties of the CD and CD novolac (CDN) cured with BMI are investigated. The curing mechanism is elucidated by the proton nuclear magnetic resonance (1H-NMR) and FTIR analyses of the model reaction products of CD and N-phenylmaleimide (PMI) and the FTIR analysis of the cured CD/ BMI (cCD/BMI) and cured CDN/BMI (cCDN/BMI) resins.

EXPERIMENTAL PROCEDURE

Materials

CD was kindly supplied from Cashew Company (Saitama, Japan). Para formaldehyde and oxalic acid dihydrate were purchased from Wako Pure Chemical Industries (Osaka, Japan) and Kanto Chemical (Tokyo, Japan), respectively. BMI was purchased from Tokyo Kasei Kogyo (Tokyo, Japan). All the commercially available reagents were used without further purification.

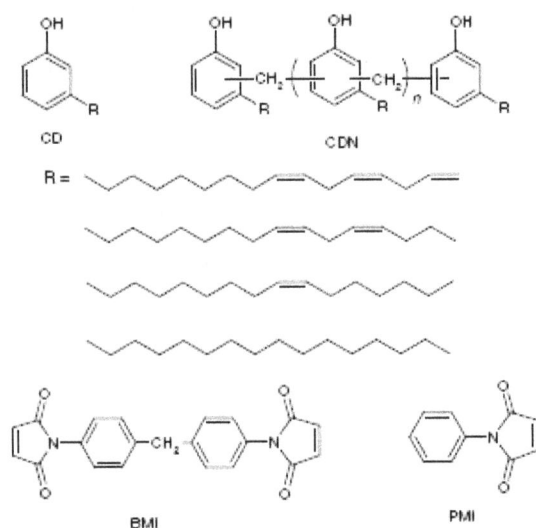

Figure 1 Structures of cardanol (CD), cardanol novolac (CDN), N-phenylmaleimide (PMI) and 4, 4o-bismaleimidediphenylmethane (BMI)

Synthesis of CDN

The molecular weight of CD (300.2), calculated from the chemical formula C21H36-2xO (average degree of unsaturation x=2.14 determined by 1H-NMR method) was used for the calculation of the feed amount of CD. A mixture of CD 30.0 g (0.100mol), paraformaldehyde 1.82 g (CH2O unit: 0.0606mol) and oxalic acid dehydrate 0.900 g (7.13mmol) was reacted at 120 1C for 6h. After chloroform (300 ml) was added to the reaction mixture, the solution was washed with water, and dried with sodium sulfate. The chloroform solution was filtered, and then concentrated in vacuo at 200 1C to produce CDN 27.3g in 89% yield.

Preparation of the cured materials of CD and BMI

A mixture of CD 13.7 g (45.6mmol) and BMI 16.3 g (maleimide unit: 91.2mmol) was stirred at 200 1C for ca. 4 h to produce a gelatinous material. The obtained compound was compression-molded at 200 1C/5MPa for 1 h, 220 1C/5MPa for 2 h and finally 250 1C/5MPa for 5 h to produce the cCD/BMI with CD/maleimide ratio of 1/2 (cCD/BMI 1/2). In a manner similar to the preparation of cCD/BMI 1/2, the cured resins of cCD/BMI with the CD/ maleimide ratios of 1/4 and 1/6 (cCD/BMI 1/4 and cCD/BMI 1/6) were also prepared.

Preparation of the cured materials of CDN and BMI

Weight per CD equivalent of CDN (307.4), which was calculated from the number of phenolic units (p=2.52) of CND evaluated from 1H-NMR analysis and the molecular weight (675.4) of CDN calculated from the chemical formula $C_{22p-1}H_{(36-2x)p}O_p$, were used for the calculation of the feed amount of CDN. The parameter, n of CDN shown in Figure 1, is expressed by the equation, n=p-2. A mixture of CDN 13.8 g (CD unit: 44.9mmol) and BMI 16.1 g (maleimide unit: 89.9mmol) was stirred at 200 1C for ca. 0.5 h to produce a gelatinous material. The compound that was obtained was compression-molded at 200 1C/5MPa for 1 h, 220 1C/5MPa for 2 h, and finally 250 1C/5MPa for 5 h to produce the cCDN/BMI with CD/maleimide ratio of 1/2 (cCDN/ BMI 1/2). In a manner similar to the preparation of cCDN/BMI 1/2, the cured resins of CDN/BMI with CD/maleimide ratios of 1/4 and 1/6 (cCDN/BMI 1/4 and cCDN/ BMI 1/6) were also prepared.

Model reactions of CD and PMI

A mixture of CD 0.928 g (3.09mmol) and PMI 1.07 g (6.18mmol) was stirred at 200 1C for 8 h to give a model reaction product of CD/PMI with CD/ maleimide unit ratio of 1/2 (cCD/PMI 1/2). The reaction mixture was used for the 1H-NMR and FTIR measurements. Similarly, cCD/PMI 1/4 and cCD/PMI 1/6 was also prepared.

Measurements

1H-NMR spectra were recorded on a Bruker AV-400 (400MHz) (Madison, WI, USA) using CDCl3 as a solvent. FTIR spectra were measured on an FTIR 8100 spectrometer (Shimadzu, Kyoto, Japan) by the KBr or attenuated total reflectance methods. The temperature at which 5% weight loss occurred was measured on a thermogravimetric analyzer TGA7 (Perkin-Elmer Japan, Yokohama, Japan) in a nitrogen atmosphere at a heating rate of 20 1Cmin_1.

The morphology of the cured resins was observed by field emission-scanning electron microscopy (FE-SEM), using a Hitachi S-4700 machine (Hitachi High-Technologies, Tokyo, Japan). All samples were fractured after immersion in liquid nitrogen for about 10 min. The fracture surfaces were sputter coated with gold to provide enhanced conductivity. Dynamic mechanical analysis of the rectangular plates (40-6-2mm3) was performed on a Rheolograph Solid instrument (Toyo Seiki, Tokyo, Japan) under air atmosphere with a chuck distance of 20mm, a frequency of 1Hz and a heating rate of 2 1Cmin-1, based on ISO 6721-4:1994 (Plastics-Determination of dynamic mechanical properties, Part 4: Tensile vibration—Non-resonance method). Flexural testing of the rectangular specimen (70-10-2mm3) was performed using an Autograph AG-1 (Shimadzu) based on the standard method for testing the flexural properties of plastics (JIS K7171: 2008 (ISO 178: 2001)). The span length was 30mm, and the testing speed was 10mmmin-1. Five specimens were tested for each set of samples, and the mean values were calculated.

RESULTS AND DISCUSSION

Characterization of CD and CDN

The 1H-NMR spectrum of the used CD in CDCl3. The 1H signals at 7.21 (m, 1H), 6.83 (m, 1H) and 6.72 p.p.m. (m, 2H) are assigned to the protons, H-b, H-c and H-a,d attached to the benzene ring of CD, respectively. As is shown in Figure 1, CD is known to contain cis, cis, cis-pentadeca-8,11,14-trienyl, cis, cispentadeca-8,11-dienyl, cis-pentadec-8-enyl and pentadecyl moieties, whose fractions are expressed by f3, f2, f1 and f0, respectively. The 1H signals at 5.10 p.p.m. (m, 0.833H) are assigned to H-l of the terminal vinyl group of the triene moiety. If it is assumed that the terminalvinyl group is contained only for the triene moiety, the fraction of the triene moiety (f3) is evaluated to be 0.417 from the equation, 0.833H/ 2H. The 1H signal due to H-k of the triene moiety was observed at 5.90 p.p.m. (m), which partly overlapped with the phenolic hydroxyl proton signal at 5.78 p.p.m. The latter hydroxy proton signal often shifted to another region and sometimes disappeared from the measurement at a different concentration. The 1H signals at 5.45 p.p.m. (m, H-i, 3.439H) are related to the cis-olefin protons of the triene, diene and monoene moieties. The degree of unsaturation (x) of CD is calculated to be 2.14 from the equation, x=3.439H/ 2Hþ0.833H/2H. Also, the following equation (1) is obtained from the equations, x=3f3+2f2+f1 and f3=0.417: 2f2 +f1 =0:889

The 1H signals of allylic position (H-h) of the triene, diene and monoene moieties were observed at 2.16 p.p.m. (bs, 3.372H), whose proton number is expressed by 2f3+4f2+4f1. Therefore, the following equation is derived: f2 +f1 =0:635

From Equations (1) and (2), f2 and f1 are calculated to be 0.254 and 0.381, respectively. The 1H signals of terminal methyl groups for the diene, monoene and pentadecyl moieties are observed at 0.98 p.p.m. (bm, 1.879H), whose proton number is expressed by 3f2+3f1+3f0.

Although f0 is calculated to be -0.009 from Equation (3), it is estimated to be 0 after considering the error of this analysis. 1:879=3f0 +3f1 +3f0 =3X0:254+3X0:381+3f0   (3)

Therefore, (f3+f2+f1+f0) is calculated to be 1.052 in this 1HNMR analysis. However, it should be 1, the values of f3, f2, f1 and f0 are compensated to be 0.397, 0.241, 0.362 and 0, respectively, after considering the error of this analysis. Their values are fairly in good agreement with the reported values (0.41, 0.22, 0.34 and 0.02) measured by chromatography.18 CDN was synthesized as a brown viscose liquid by the reaction of CD and paraformaldehyde in the presence of oxalic acid at the feed molar ratio of CD/CH2O 1.65/1. Theoretical number of phenolic units (p) is 2.54 based on the equation, p/(p-1)=1.65 for the addition–condensation reaction. Figure 2b shows the 1H-NMR spectrum of CDN in CDCl3. Benzylic methylene protons (Ph-CH2- Ph) and aromatic protons (Ar-H) were observed at 3.88 p.p.m. (m) and 7.2–6.6 p.p.m. (m), respectively.

In this case, phenolic hydroxyl protons were not observed. From the ratio of [Ph-CH2-Ph]/ [Ar-H]=(2p-2)/(2p+2)=0.4326, p is calculated to be 2.52, which is in good agreement with the theoretical value (2.54). As the number of unsaturated carbon chains of CDN is more than two, cross linked structure can be formed even if the CD unit reacts with maleimide unit of BMI at the ratio of 1/1. In the case of the reaction of CD and BMI, the polymer network would not be formed if the stoichiometric ratio of maleimide/CD was 1/1.

Model Reaction of CD and PMI

The model reactions of CD/PMI with molar ratios of 1/2, 1/4 and 1/6 at 2001C for 8 h were carried out to elucidate the curing mechanism of cCD/BMI or cCDN/ BMI. The 1H-NMR spectra of cCD/PMI 1/2 and cCD/PMI 1/6 in CDCl3. The olefinic protons of PMI at 6.85 p.p.m. (s) completely disappeared for cCD/PMI 1/2, and new 1H signals were observed at the regions of 6.2–5.2 p.p.m. (H-k) and 3.7–2.7 p.p.m. (H-j). The former proton signals are reasonably assigned to olefinic protons, and the latter ones to methylene or methine protons adjacent to carbonyl or vinyl groups. The 1H signals due to the original cis-olefin moiety (H-i) decreased with increasing amount of PMI. In the case of CD/PMI 1/6, the facts that the 1H signals (H-i) was very weak and that broad aliphatic 1H signals appeared at lower than 3.7 p.p.m. increased as is shown in Figure 3b, indicating the occurrence of addition copolymerization of the original cis-olefin moieties and PMI. Also, the appearance of broad 1H signals around 4 p.p.m. suggests the occurrence of homopolymerization of PMI.27 However, the fact that the olefinic proton signal of maleimide moieties is observed at 6.85 p.p.m. for cCD/PMI 1/6 indicates that excess PMI remains at the reaction temperature of 200 1C. The probable reaction mechanism of CD and PMI elucidated from the 1H-NMR.

Thus, the end reaction of the terminal allyl group of the pentadeca-8, 11,14-triene moiety and PMI generates the cis, cis, trans-pentadeca-8,11,13-triene moiety. Subsequently, the formed conjugated diene moiety undergoes a Diels–

Alder reaction with PMI to produce cyclohexene moiety. When the trans-olefin generated by the ene reaction of isolated cis-olefin and PMI is not conjugated with another cis-olefin, the subsequent Diels–Alder reaction does not occur. It is also considered that addition copolymerization of the olefins originated from CD and PMI and the homo polymerization of PMI occur in addition to the ene and Diels–Alder reactions.

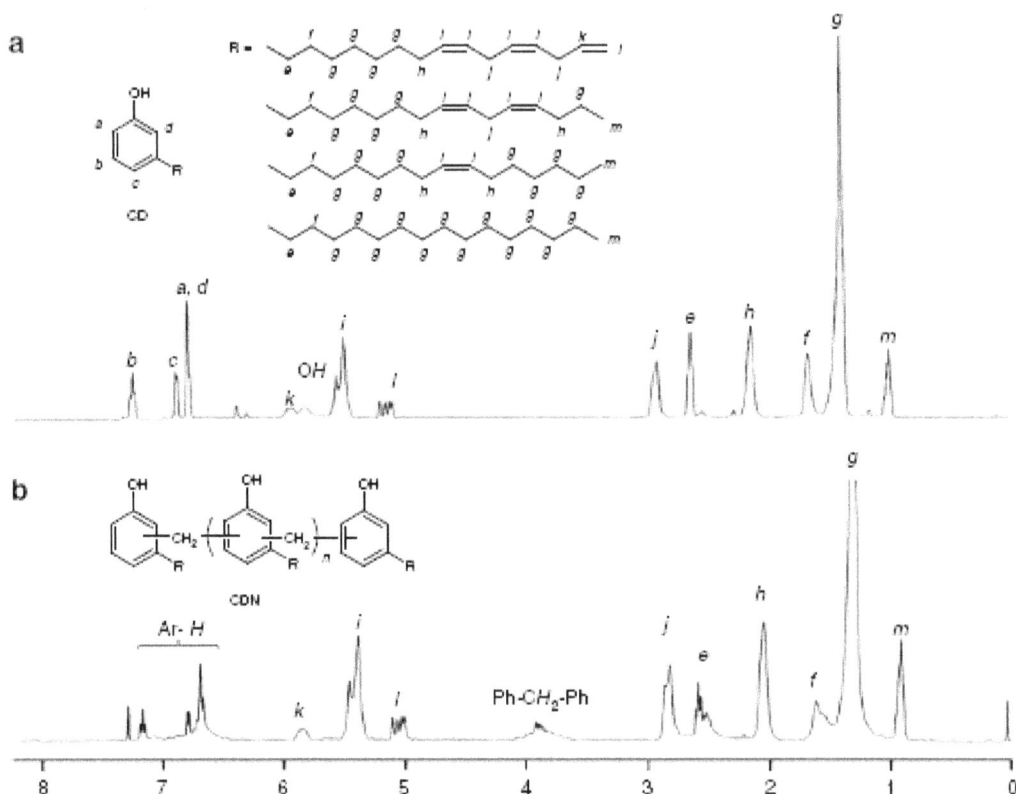

Figure 2 Proton nuclear magnetic resonance (1H-NMR) spectra of (a) cardanol (CD) and (b) cardanol novolac (CDN) in CDCl3.

The FTIR spectra of the model reaction products were analyzed to verify the reaction mechanism elucidated by 1H-NMR for cCD/PMI. The analysis of the C–H out-of-plane bending absorptions of alkenes is effective to identify their substitution manner and cis–trans isomerism. For example, it is known that cis-disubstituted RHC=CHR and trans-disubstituted RHC=CRH show the absorption peaks at 730–665 cm_1 (strong) and 980–960cm_1 (strong), respectively. It is also known that vinyl RHC¼CH2 shows the absorption peaks at 995–985 cm_1 (strong) and 915–905 cm_1 (strong). The C–H out-of-plane bending absorption peak of cisdisubstituted RHC¼CHR for CD was observed at 719cm_1. Also, the C–H out-of-plane bending absorption peaks of the terminal vinyl group for CD were observed at 991 and 910 cm_1.

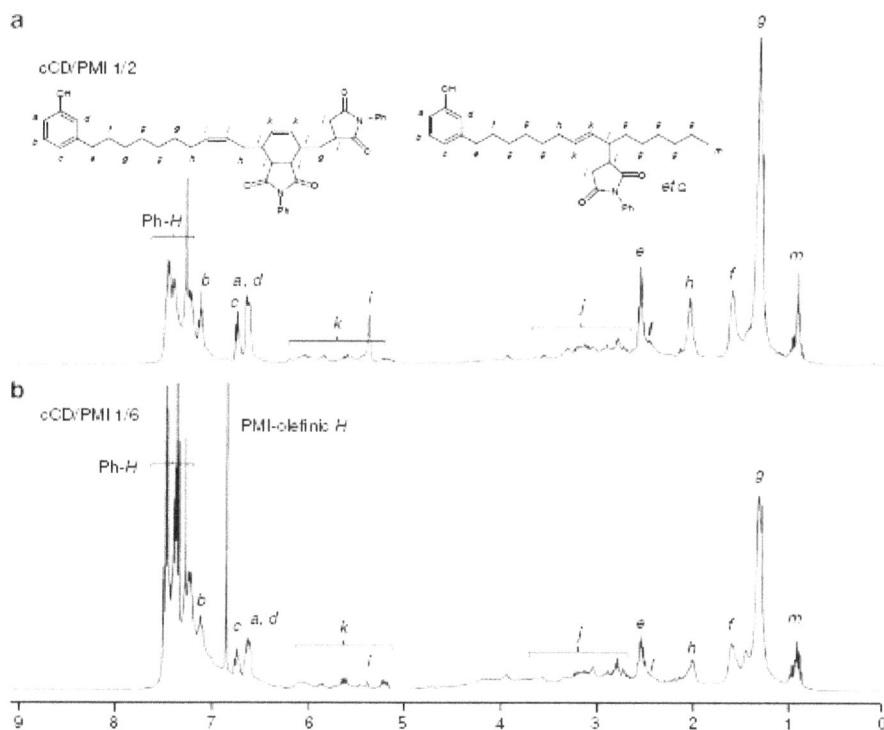

Figure 3 Proton nuclear magnetic resonance (1H-NMR) spectra of (a) cured cardanol (cCD)/N-phenylmaleimide (PMI) 1/2 and (b) cCD/PMI 1/6 in CDCl3.

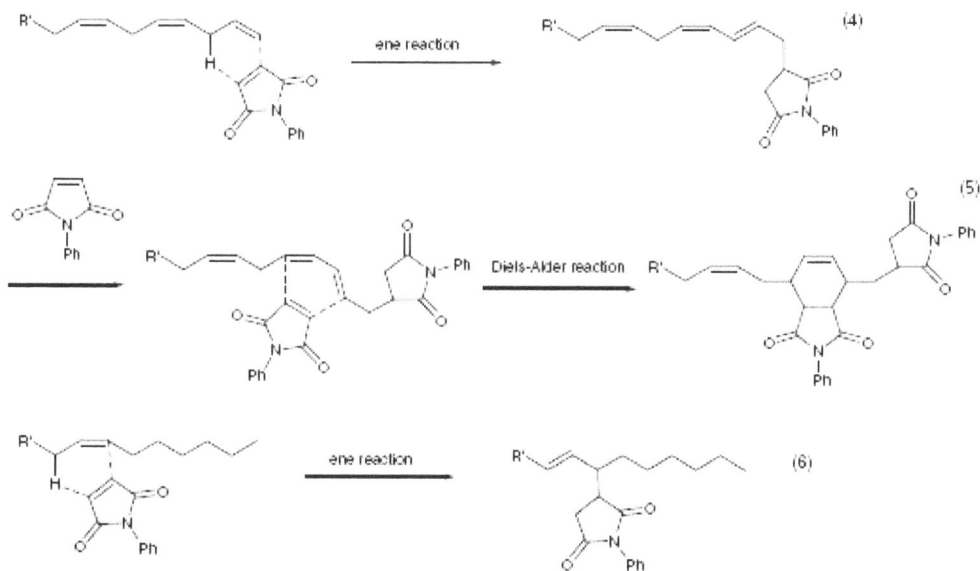

Figure 4 Probable reaction mechanism of cardanol (CD) and N-phenylmaleimide (PMI).

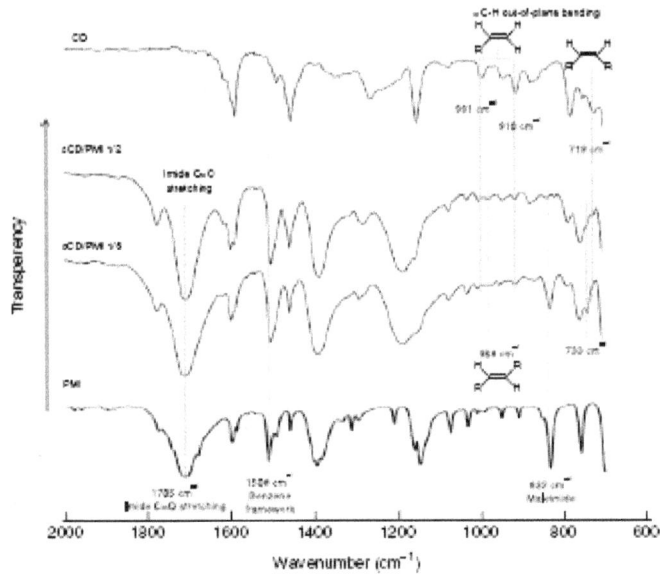

Figure 5 Fourier transform infrared (FTIR) spectra of cardanol (CD), cured CD (cCD)/N-phenylmaleimide (PMI) 1/2, cCD/PMI 1/6 and PMI.

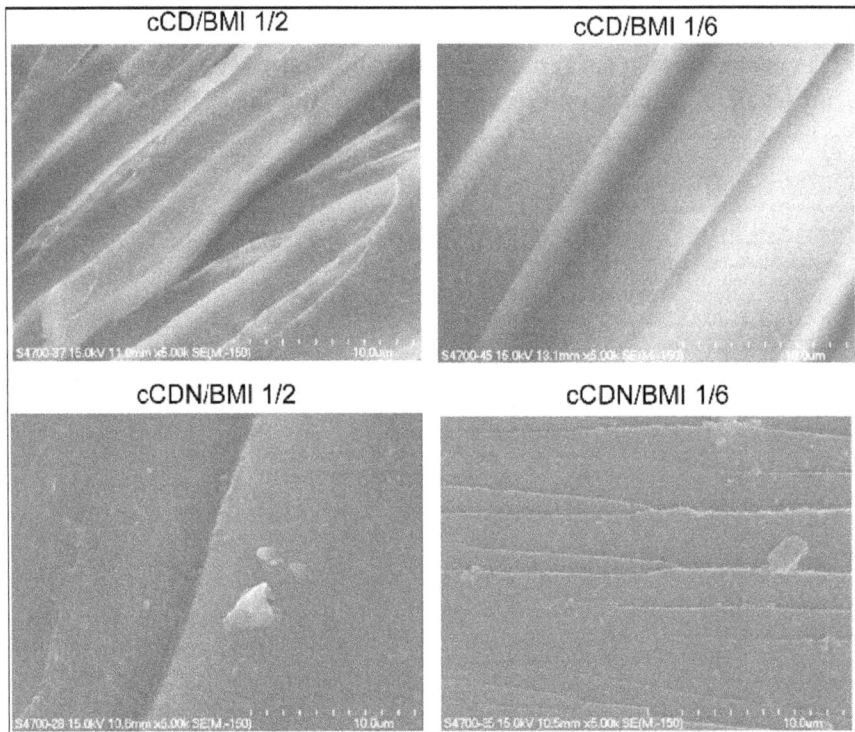

Figure 6 Field emission-scanning electron microscopy (FE-SEM) images of the fractured surfaces of cured cardanol (cCD)/4, 40-bismaleimidediphenylmethane (BMI) and cured CD novolac (cCDN)/BMI resins.

In the FTIR spectra of cCD/PMI 1/2 and cCD/PMI 1/6, the absorption peaks related to cis-RHC=CHR and vinyl RHC=CH2 observed for CD considerably diminished, and new absorption peaks around 968 cm_1 related to the trans-RHC=CRH unit formed by the ene reaction appeared. Also, another new absorption peak at 733cm_1 related to cis-RHC=CHR of the cyclohexene moiety formed by the ene reaction and subsequent Diels–Alder reaction. Although the absorption peak at 832 cm_1 related maleimide group disappeared for cCD/ PMI 1/2, this peak was observed for cCD/PMI 1/6. The result obtained from the FTIR analysis is in good agreement with that elucidated from the 1H-NMR analysis.

Curing reactions of CD/BMI and CDN/BMI

The mixture of CD and BMI with the unit ratio of CD/maleimide 1/2–1/6 gelled during the prepolymerization at 200 1C for 4 h, indicating the progress of crosslinking reaction. Similarly, the mixture of CDN and BMI gelled during the pre-polymerization at 200 1C for 0.5 h. The obtained compounds of CD/BMI and CDN/BMI were compression-molded finally at 250 1C for 5 h. FE-SEM images of the fractured surface of cCD/BMIs and cCDN/BMIs. All the cured resins of CD/BMI and CDN/ BMI did not show any phase separation, indicating that CD and CDN copolymerized with BMI. To verify the curing mechanism estimated from the model reaction, FTIR analysis of cCDN/BMI was performed. Figure 7 shows the FTIR spectra of CDN, cCDN/BMI 1/2, cCDN/BMI 1/6 and BMI over the range between 2000 and 700cm_1. The C–H out-of-plane bending absorption peak of cis-disubstituted RHC=CHR was observed at 719cm_1 for CDN in a manner similar to CD. Also, the C–H out-of plane bending absorption peaks of terminal vinyl group for CDN were observed at 991 and 910cm_1. The absorption peaks related to the cis-RHC=CHR and vinyl RHC=CH2 observed for CDN almost disappeared for cCDN/BMI 1/2 and cCDN/BMI 1/6 and new broad absorption peaks around 960–980 cm_1 related to trans-RHC=CRH were observed, indicating the occurrence of the ene reaction.

Although the absorption peak related to the cyclohexene moiety generated by the Diels–Alder reaction was clearly observed at 733 cm_1 for cCD/PMI, the peak at a near wave number region (738 cm_1) was very weak for cCDN/BMI 1/2 and cCDN/ BMI 1/6. There is a possibility that retro-Diels–Alder reaction and subsequent addition copolymerization occurred at the curing temperature of 250 1C, or that Diels–Alder reaction itself is prevented by a steric hindrance by use of BMI instead of PMI and the olefinic moieties of CDN directly copolymerized with BMI.

The comparison of FTIR spectra of CD/PMI 1/4 and CDN/BMI 1/4 pre-polymerized at 200 1C for 8 h. In contrast to the fact that the pre-polymerized CD/PMI showed a clear peak at 733 cm_1 based on a cyclohexene moiety, the pre-polymerized CDN/ BMI showed no absorption peak at a similar wave number region, except for the peaks based on the unreacted BMI. A similar trend was observed for CDN/BMI 1/2 and CDN/BMI 1/4 pre-polymerized at 200 1C for 8 h. These results strongly support that the Diels–Alder reaction is highly hindered in the case of the reaction of CDN and BMI. Although we did not show the FTIR spectra of cCD/BMI 1/2 and cCD/BMI 1/6, their spectra were very similar to those of cCDN/BMI 1/2 and cCDN/BMI 1/6, respectively, indicating that the ene reaction and addition copolymerization similarly occurred for cCD/BMIs. Regarding the effect of steric hindrance on Diels–Alder reaction, it is reported that the Diels– Alder reaction did not occur but the ene reaction and subsequent addition copolymerization occurred in the reaction of DABA/PMI or DABA/BMI, whereas the model reaction of 2-allylphenol and PMI produced the ene and subsequent Diels–Alder adducts.

Thermal and mechanical properties of cCD/BMI and cCDN/BMI dynamic mechanical analysis curves of cCD/ BMIs and cCDN/BMIs cured at 250 1C, respectively. The cCD/BMI 1/ 2 and cCDN/BMI 1/2 showed clear tan d peak temperature corresponding to glass transition temperature (Tg) at 141.7 and 158.0 1C, respectively. The storage modulus (E0) for cCD/BMI 1/2 and cCDN/ BMI 1/2 considerably dropped at around the Tg. However, clear glass transition and decline of E0 were not identified until 300–350 1C for the cured resins with a higher BMI content than cCD/BMI 1/2 and cCDN/BMI 1/2, indicating highly cross linked maleimide resins with superior heat resistance are obtained.

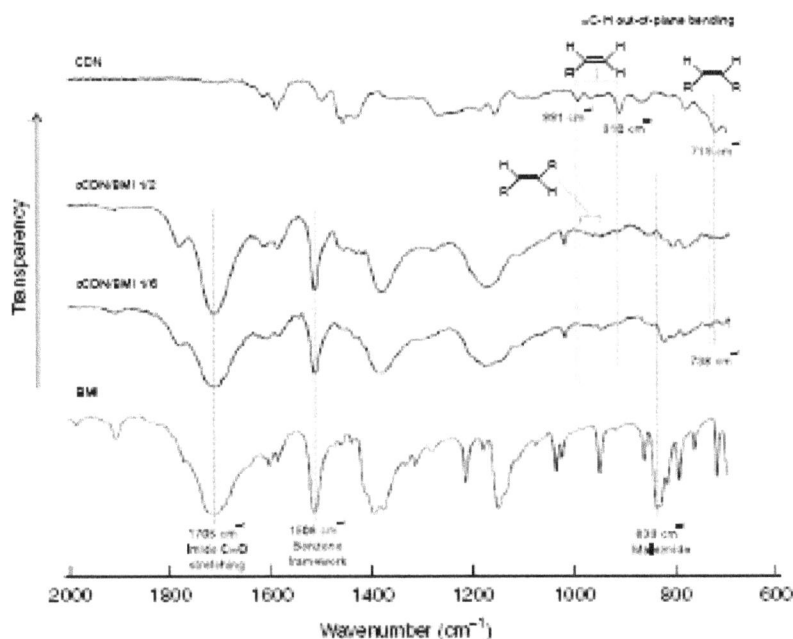

Figure 7 Fourier transform infrared (FTIR) spectra of cardanol novolac (CDN), cured CDN (cCDN)/4,40-bismaleimidediphenylmethane (BMI) 1/2, cCDN/BMI 1/6 and BMI.

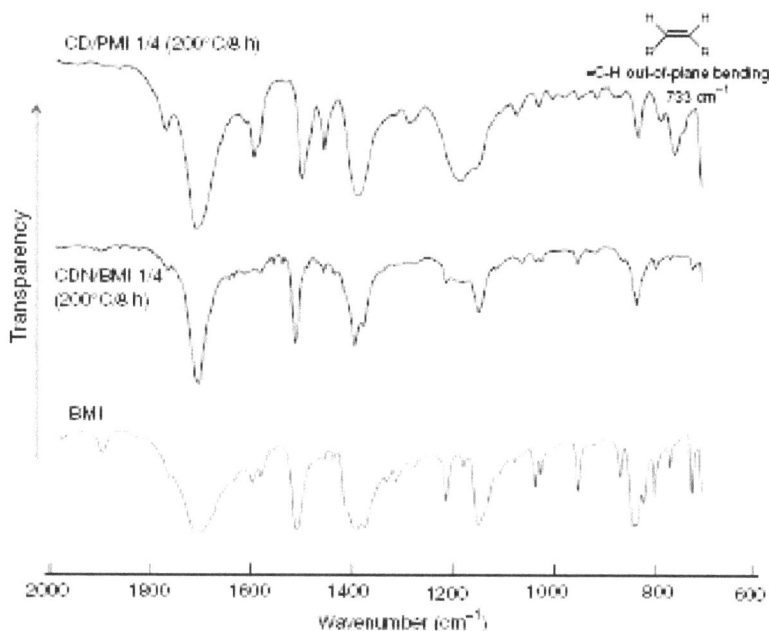

Figure 8 Fourier transform infrared (FTIR) spectra of the compounds of cardanol (CD)/N-phenylmaleimide (PMI) 1/4 and CD novolac (CDN)/4,40-bismaleimidediphenyl methane (BMI) 1/4 pre-polymerized at 2001C for 8 h and BMI.

Thermogravimetric analysis curves of cCD/BMI and cCDN/BMI cured resins. The 5% weight loss temperature increased with increasing BMI content, and all the cured resins except cCD/BMI 1/2 showed an excellent 5% weight loss temperature higher than 450 1C. When the cured resins with the same unit ratio of CD /maleimide are compared, cCDN/BMI had a higher 5% weight loss temperature than cCD/BMI did.

Flexural properties of cCD/BMIs and cCDN/BMIs. There were little differences in flexural strength and modulus between the cCD/BMI and cCDN/BMI with the same unit ratio, suggesting that polymer network is formed for both the cured resins by the reaction at the stoichiometric ratio of maleimide/CD more than two. When the unit ratio is 1/4, the highest flexural strength and modulus were attained for both the resin systems, except that cCDN/BMI =showed a lower flexural strength than cCDN/BMI 1/2 did. These results suggest that the introduction of a succinimide moiety with a higher fraction during the addition copolymerization contributes to the improvement of the thermal and flexural properties. The fact that flexural strength and modulus of cCD/BMI 1/6 and cCDN/BMI 1/6 were lower than those of the corresponding 1/4 samples may be attributed to the brittle character of homopolymerized BMI component and/or some delamination between succinimide-rich and –poor components.

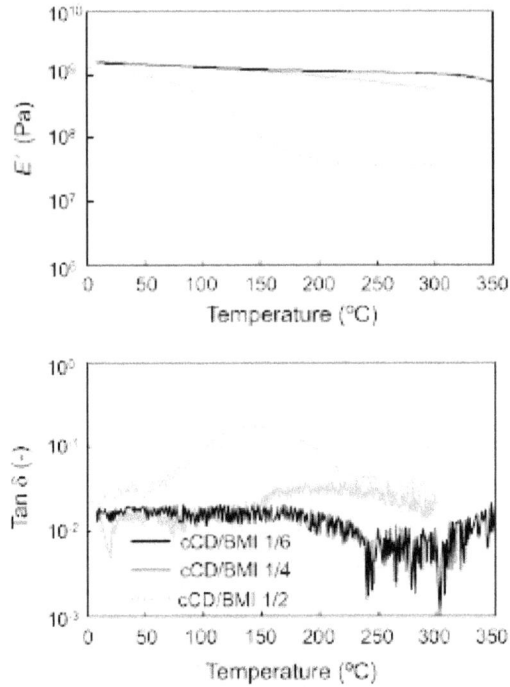

Figure 9 Dynamic mechanical analysis (DMA) curves of cured cardanol (cCD)/4,40-bismaleimidediphenyl methane (BMI) 1/2, cCD/BMI 1/4 and cCD/ BMI 1/6.

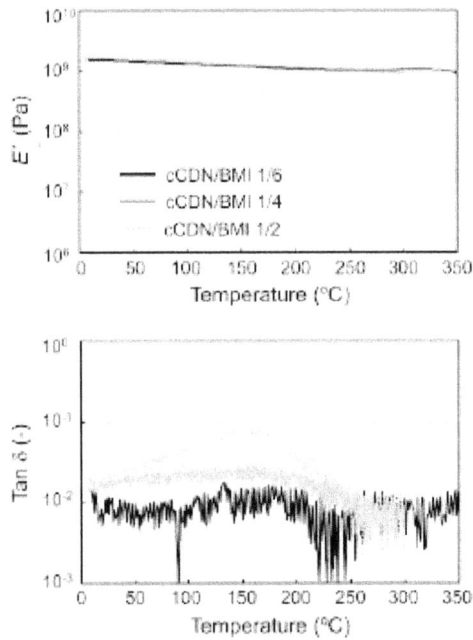

Figure 10 Dynamic mechanical analysis (DMA) curves of cured cardanol novolac (cCDN)/4,40-bismaleimidediphenyl methane (BMI) 1/2, cCDN/BMI 1/4 and cCDN/ BMI 1/6.

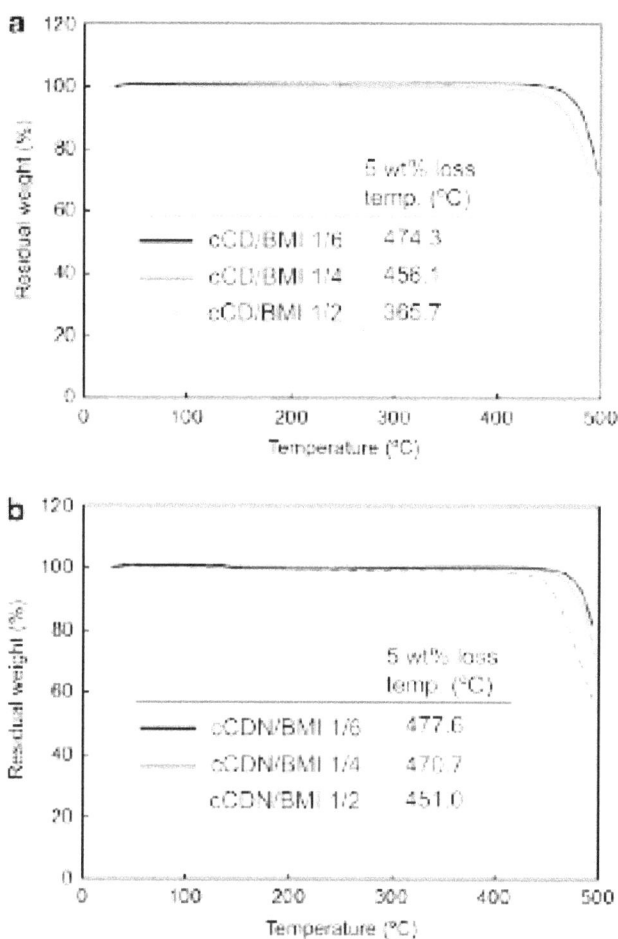

Figure 11 Thermogravimetric analysis (TGA) curves of (a) cured cardanol (cCD)/4,40-bismaleimidediphenyl methane (BMI) and (b) cured CD novolac (cCDN)/BMI resins.

CONCLUSION

CDN (p=2.52) was synthesized by the reaction of CD (x=2.14) and par formaldehyde. The pre-polymerized compounds of CDN/BMI and CD/BMI with CD/maleimide ratios 1/2, 1/4 and 1/6 at 200 1C were finally compression-molded at 250 1C for 5 h to produce cCDN/ BMI and cCD/BMI, respectively. The FE-SEM analysis revealed that homogeneous cured resins without phase separation were obtained for cCD/ BMI and cCDN/BMI. Although the 1H-NMR and FTIR analyses of the model reaction product of CD and PMI at 200 1C for 8 h suggested the occurrence of the ene reaction and subsequent Diels–Alder reaction, the FTIR analysis of cCD/BMI and cCDN/BMI suggested the occurrence of the ene reaction and addition copolymerization. The cCDN/BMI and cCD/BMI with CD/maleimide ratio lower than 1/2 did not show glass transition until 300 1C and had a 5% weight loss temperature higher than 450

1C. The cCDN/BMI and cCD/BMI with CD/maleimide ratio 1/4 showed the most balanced flexural properties (flexural strength 60–80MPa, flexural modulus 2.0–2.5 GPa).

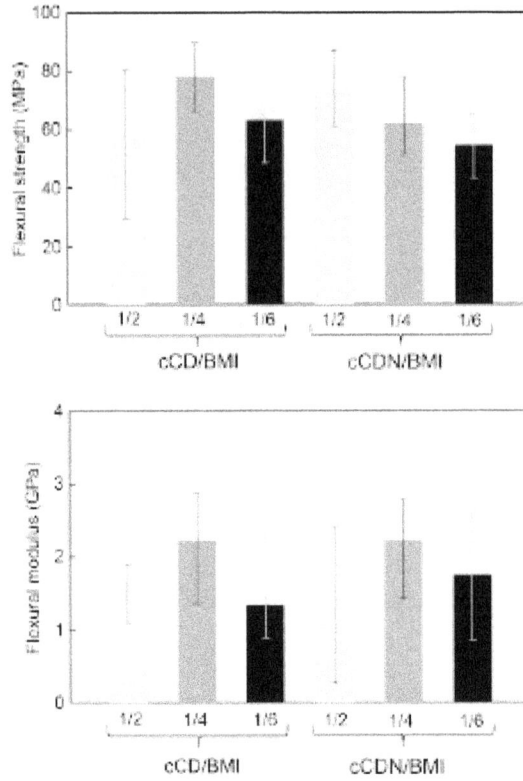

Figure 12 Flexural properties of cured cardanol (cCD)/4,40-bismaleimidediphenyl methane (BMI) and cured CD novolac (cCDN)/BMI resins.

5
Chapter

CASTOR OIL AS A POTENTIAL RENEWABLE RESOURCE FOR THE PRODUCTION OF BIO BASED POLYMER

INTRODUCTION

Castor plant (*Ricinus communis*) is from the family Euphorbiaceae and grows wild in varied climatic conditions. The plant produces castor seeds that contain up to 50 % castor oil by weight. The oil can easily be extracted from castor seeds and find its use in a multitude of sectors such as medicine, chemicals industry and in other technologies. The demand for castor oil and its products in the world market has been on the steady increase partly due to their renewable nature, non-competition with food, biodegradability, low costs, and eco-friendliness.

It is now estimated that the oil has over 700 industrial uses and the uses keeps on increasing. The chemistry of castor oil is mainly centered on ricinoleic acid due to its high content in the oil and the presence of the three functional groups in the acid. The three functionalities are crucial towards the versatility of the oil for the production of variety of castor oil based products. The carboxylic group for instance, can lead to a wide range of esterification products while the single point of instauration can be altered by hydrogenation, epoxidation or vulcanization. On the other hand, the hydroxyl functional group at carbon-12, can be acetylated, alkoxylated or removed by dehydration to increase the instauration of the oil.

The reactions of castor oil are becoming of high industrial importance. This paper reviews on the geographical distribution of castor plants and the world production of castor seeds and castor oil. Furthermore, some important reactions on converting castor oil into useful products are discussed. The reactions discussed include hydrogenation, pyrolysis, caustic fusion, dehydration, transesterification, sulphonation, and polymerization. The use of castor oil and ricinoleic acid as green capping agent in the synthesis of nonmaterial's is highlighted.

CASTOR OIL AND DERIVATIVES GLOBAL MARKET

Owing to its rich properties and variety of end-uses, together with increased interests in biopolymer and biofuels industries, the potential for castor oil to play a much larger role in the world economy has increased dramatically in recent years. For instance, trend shows that castor oil prices steadily rose from $946 per tone in 2002–2003 to $2390 in 2010–2011. The contribution of castor oil on the world economy is expected to continue increasing and it has been predicted that the global castor oil and derivatives market will reach USD 1.81 billion by 2020. The increased interest of substitution of conventional fuel by bio fuels, volatile crude oil prices, higher demand from Europe, China and the US, and growth of key endues industries including cosmetics and lubricants are expected to drive the global castor oil and derivatives market. On the other hand, threat from other vegetable oils in terms of price and application, and high dependency on seasonality may hinder the market growth. Some major companies operating in the castor oil and derivatives global market are: Thai Castor Oil Industries Co.

EXTRACTION, COMPOSITION AND PROPERTIES OF CASTOR OIL

Castor oil can be extracted from castor seeds by solvent extraction, mechanical pressing or a combination of both. Mechanical pressing is disadvantageous as it can only extract about 45 % of the oil.

Fig. 1 Castor plant (a) and castor seeds (b)

This means that the rest of the oil in the cake must again be extracted by a solvent and therefore causing double work, increase in extraction expenses and thus the process become environmentally unfriendly. On average, castor seeds contain between 45 and 55 % oil by weight depending on the varieties, geographical location and the method of extraction.

Like other vegetable oils, castor oil exists as a mixture of saturated and unsaturated fatty acids attached to a glycerol. In the mixture of castor oils fatty acids, ricinoleic acid accounts for about 90 % of the mixture with other components in small proportions of not more than 5 %.

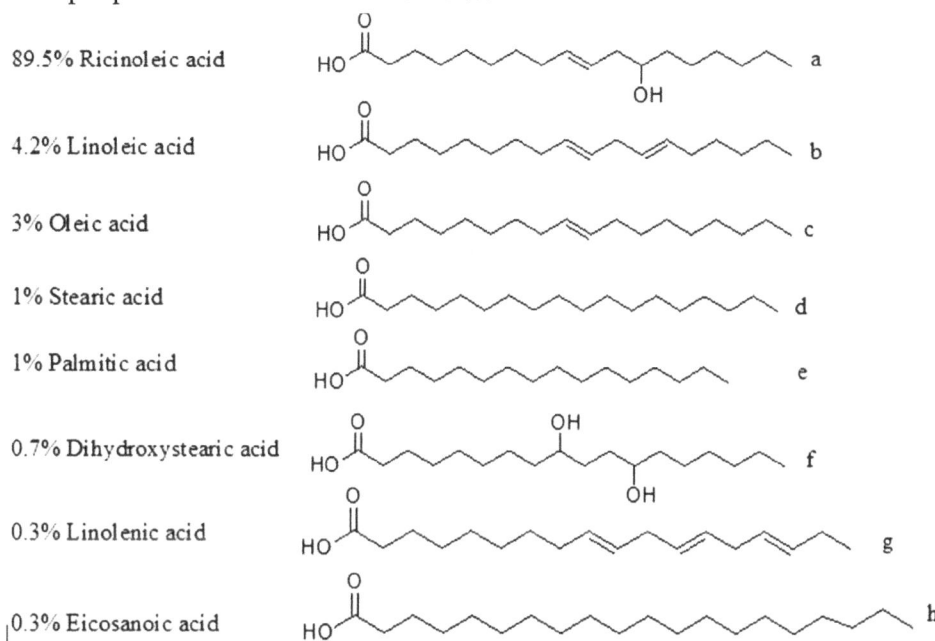

89.5% Ricinoleic acid a
4.2% Linoleic acid b
3% Oleic acid c
1% Stearic acid d
1% Palmitic acid e
0.7% Dihydroxystearic acid f
0.3% Linolenic acid g
0.3% Eicosanoic acid h

Fig. 2 Composition of castor oil fatty acids

Castor oil is extracted colourless to very pale yellow viscous liquid with a distinct taste, mild odour and it boils at 586 K. The hydroxyl group in ricinoleic acid account for the unique properties of castor oil. For instance, the oil has relatively high viscosity and specific gravity; it is soluble in alcohols in any proportion and has limited solubility in aliphatic petroleum solvents. In addition, the polar hydroxyl group in castor oil makes it compatible with plasticizers of a wide variety of natural and synthetic resins, waxes, polymers and elastomers. Notable changes on the properties of the castor oil can also be due to several factors such as the method of extraction, seed varieties, weather conditions and soil type. For instance, cold-pressed castor oils have low acid value, low iodine value and a slightly higher saponification value than solvent-extracted oil. It has further been observed that castor seeds from different climatic conditions produce castor oils of different composition and physical–chemical properties.

Malaysian castor seeds for instance, contain total lipids (castor oil) reaching up to 43.3 % per dry weight and a saponification value of 182.96 mg KOH/g while for the Nigerian castor seeds, the total lipids (castor oil) is 48 % per dry weight with a saponification value of 178.00 mg KOH/g .

Chemical transformations of castor oil

The unique properties and diverse applications of castor oil and its derivatives make castor oil popular and even more important among vegetable oils. The presence of ester linkage, a double bond and the hydroxyl group in ricinoleic acid favours the oil as a suitable renewable resource for many chemical reactions, modifications and transformations.

The presence of carboxylic group for example, allows transformation of castor oil via several reactions such as esterification, amidation whereas the presence of a double bond, affords the transformation of the oil through reactions such as hydrogenation, carbonylation and epoxidation.

Furthermore, the hydroxyl functional group can be acetylated alkoxylated or removed by dehydration to increase the unsaturation of the oil. Catalytic dehydration leads into formation of a new double bond in the chain of ricinoleic acid resulting into a conjugated acid. This change imparts good flexibility, rapid drying, excellent color retention, and water resistance for protective coatings. Both ring-opened glyceryl ricinoleates and epoxy alkyl ricinoleates functionalized castor oil derivatives have recently been prepared with very high yields. The ring-opened glyceryl ricinoleates was achieved through catalytic ring opening and transesterification using epoxidized castor oil (ECO) as a raw material using Amberlyst 15 acid catalyst while the epoxy alkyl ricinoleates was achieved by transesterification of ECO with methanol using CaAl-layered double hydroxide base catalyst. Interestingly, the physical properties of these functionalized castor-based derivatives further demostrate the opportunity to design tailor-made materials suiting industrial needs from the oil.

Pyrolysis of castor oil cleaves the molecule to produce new useful compounds such as undecylenic acid and heptaldehyde. Addition of hydrogen bromide to the cleaved castor oil produces 11-bromo undecanoic, which upon reacting with ammonia, forms 11-aminoundecanoic acid; a monomer for nylon 11 polymer. Generally, the three functional groups in ricinoleic acid provide multitude of possibilities of converting or modifying castor oil into many other useful products depending on the intended specific uses. Chemical transformations of castor oil into castor oil based products are discussed in the subsequent sections.

Hydrogenation

Addition of hydrogen to the unsaturated fatty acid in the presence of nickel or palladium catalyst transforms the liquid ricinoleic acid into semi-solid saturated 12-hydroxystearic acid. (Scheme 1)

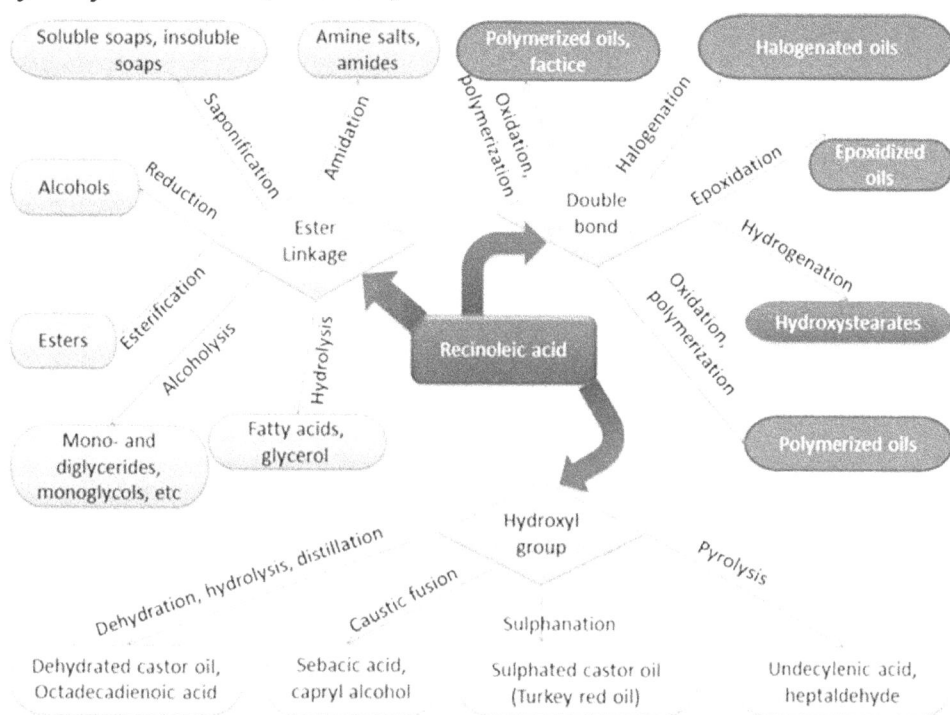

Fig. 3 Some potential ricinoleic acid reactions of industrial importance

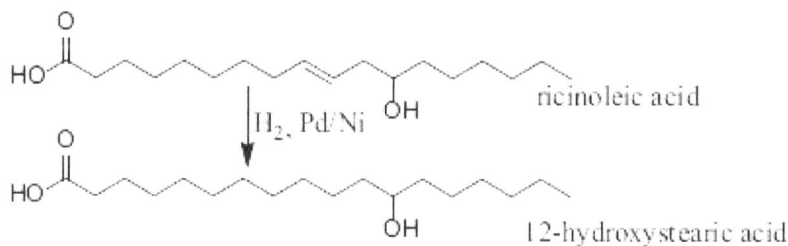

Scheme 1 Hydrogenation of ricinoleic acid

The semi-solid saturated ricinoleic acid is a valuable material in industries and in resin or polymer mixtures. The oil has high melting point, improved storage qualities, taste, and odor. Moreover, the hydrogenated oil has an improved oxidative and thermal stability. A good quality hydrogenated castor with high hydroxyl value and low iodine value is obtained at 423 K; 1.034×106 Pa; in 5 h with 2 % (weight of oil) Raney nickel catalyst . Hydrogenation of castor oil at low pressure ($1.96–2.45 \times 105$ Pa) and low temperature (398–408 K) requires high catalyst concentration.

Castor oil hydrogenation can also be done by catalytic transfer hydrogenation (CTH) (Scheme 2). Catalytic transfer hydrogenation has the advantage in that it can utilize organic molecules as hydrogen donors at ambient pressure and moderate temperatures. Moreover, no special reactors are required and the solvent is used as a hydrogen donor along with a selected catalyst. The CTH process using Pd/C catalyst and different hydrogen donor solvents have also been reported for soy, sunflower, and castor oils.

Hydrogenated castor oil (HCO) is insoluble in water and in most organic solvents but it is soluble in hot organic solvents like ether and chloroform. This insolubility is among good qualities that make HCO valuable for lubricant industries because of water resistance and retention of its lubricity.

Moreover, the polarity and surface wetting properties of HCO are useful in cosmetics, hair dressing, and solid lubricant, and paint additives, manufacture of waxes, polishes, carbon paper, candles and crayons.

Pyrolysis

Pyrolysis is an increasingly popular option for convertin biomass to solid, liquid, and gaseous fuels. It is a thermal treatment of a biomass in the absence of air that decomposes organic biomass into low molecular weight liquid, solid and gaseous products. The absence of air during pyrolysis prevents the combustion of biomass into carbon dioxide. Pyrolysis is normally done at medium to high temperature (623–1023 K) in which the biomass is degraded to yield pyrolysis oil or bio-oil. An extensive review on pyrolysis of different vegetable oils such as tung oil, sunflower oil, canola oil, soybean oil, palm oil, macauba fruit oil, cooking oil, palm oils, soybean, and castor oils have been reported . Generally, the process can be done by either direct thermal cracking or by a combination of thermal and catalytic cracking.

Reaction

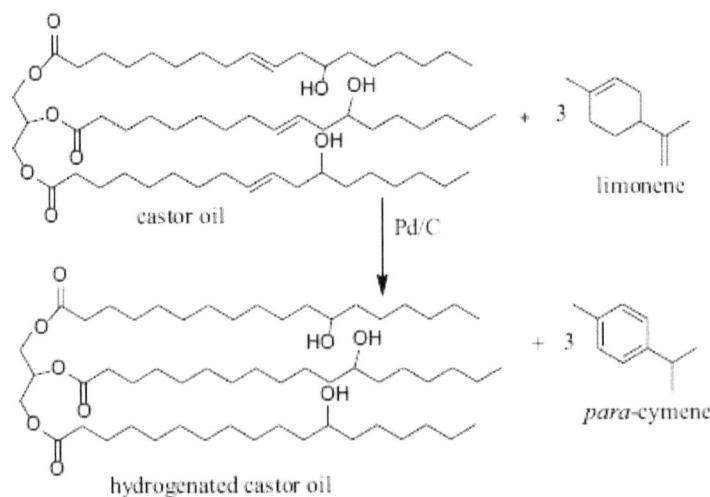

Scheme 2 Catalytic transfer hydrogenation of castor oil

Products depend on the catalyst type and the reaction conditions and can range from diesel like to gasoline like fractions. Pyrolysis of triglycerides represent an alternative method for producing renewable bio-based products suitable for use in fuel and chemical applications.

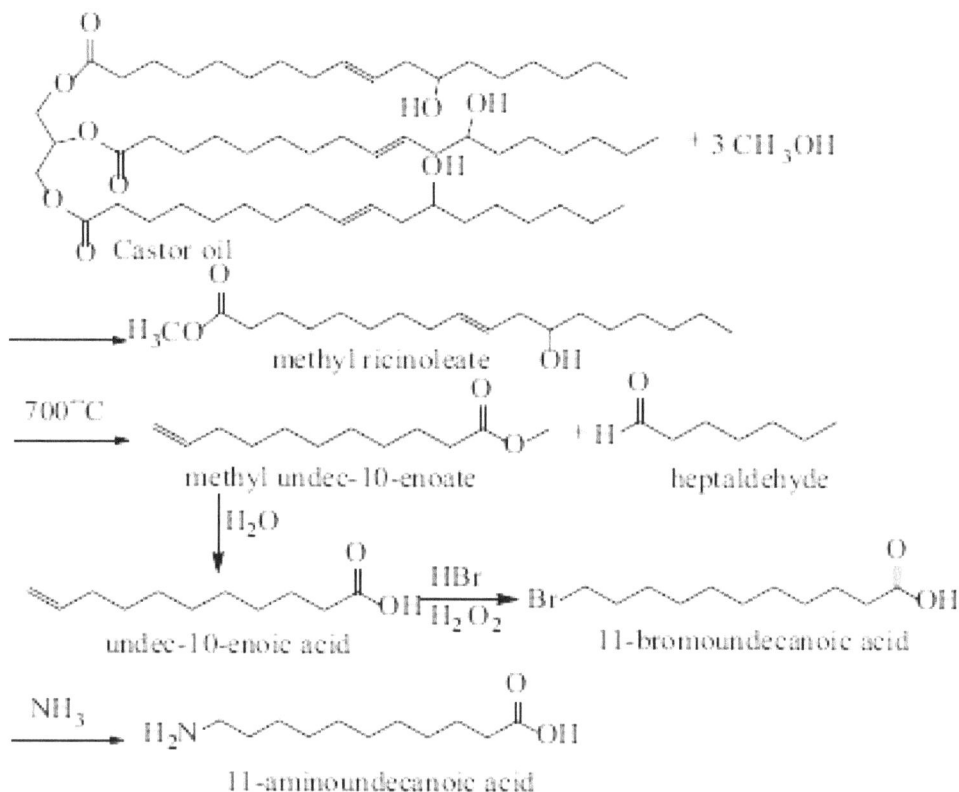

Scheme 3 Conversion of ricinoleic acid into monomers for nylon 11

Hydrolysis of castor oil

Pyrolysis of castor oil for example, at 623 K with 20 min residence time and 1.47–1.96 × 106 Pa initial hydrogen pressure, produces castor bio-oil. Moreover, methanolysis of castor oil yields methyl ricinoleate which upon pyrolysis under reduced pressure (6.0 × 103 Pa), at about 973 K, produces heptaldehyde and undecylenic acid (Scheme 2). The two products are vital intermediates in the perfumery, pharmaceutical and polymeric formulations.

Heptaldehyde is also an organic solvent for various polymers and a source of emulsifier, plasticizers and insecticides. For the undecylenic acid, it serves as a source of bactericides and fungicides but also further reactions on the acid (Scheme 3) can produce monomers for the formation of nylon 11.

Hydrolysis of castor oil by slow addition of castor oil to 80 % caustic solution (sodium hydroxide) produces ricinoleic acid and glycerol. Upon heating at 523 K in the presence of NaOH, sebacic acid (a 10 carbon dicarboxylic acid) and capryl alcohol (2-octanol) are produced (Scheme 4).

Both sebacic acid and capryl alcohol have many uses. The alcohol finds its uses as plasticizer, as a solvent, dehydrater, antibubbling agent and also as a floatation agent in coal industry. The esters of sebacic acid on the other hand are plasticizers for vinyl resins and are also used in the manufacture of dioctyl sebacate (DOS), a jet lubricant and lubricant in air cooled combustion motors Furthermore; sebacic acid is used as a monomer where it reacts with hexamethylenediamine to produce nylon 6–10.

Dehydration

The dehydration of ricinoleic acid is an acid catalysed reaction which removes the hydroxyl group in the form of water to introduce a new double bond. The reaction results into the production of both non-conjugated linoleic acid and the conjugated linoleic acid (Scheme 5).

The dehydration of castor oil is usually done at temperatures above 473 K in the presence of an acid catalyst such as concentrated sulphuric acid, phosphoric acid, p-toluenesulfonic acid, sodium bisulfate, or activated clays under inert atmosphere. The formed linoleic acids have various industrial applications including the production of protective coating, vanishes, lubricants, soaps, paints, inks, manufacture of alkyd resins, coatings, appliance finishes and primers. The linoleic acids are also basic ingredients in racing motor oil for high-performance automobile motorcycle engines.

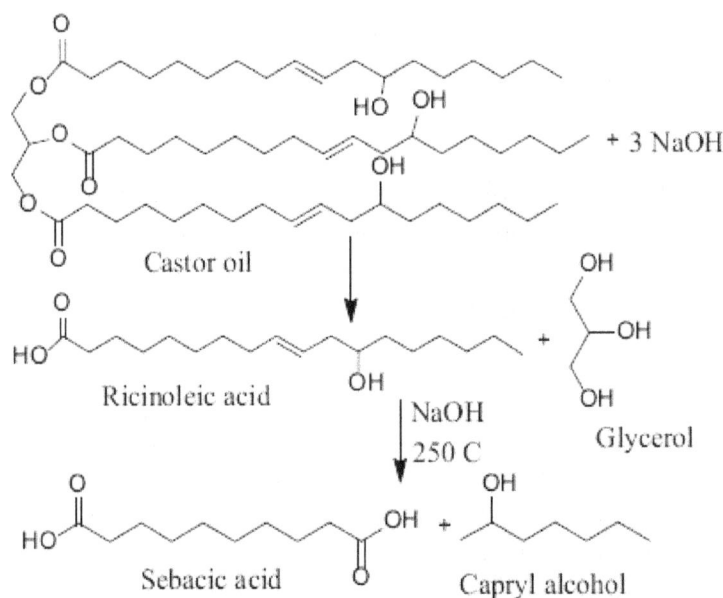

Scheme 4 Conversion of ricinoleic acid into sebacic acid and capryl Alcohol

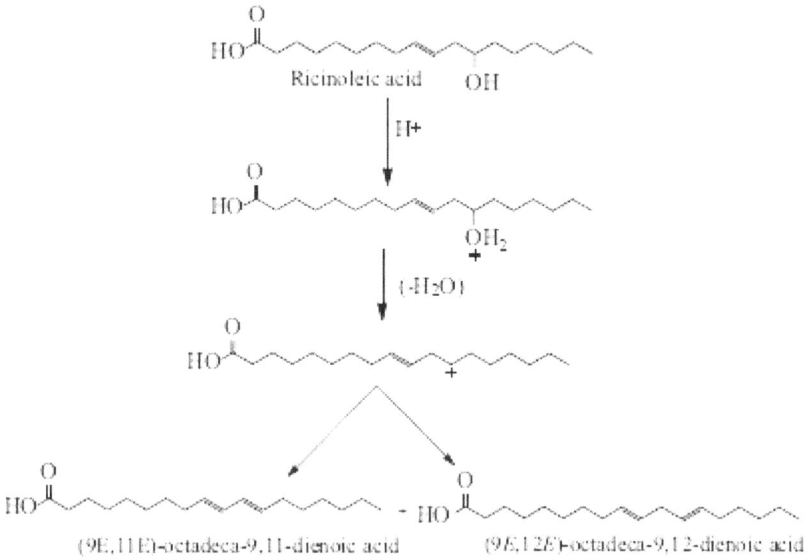

Scheme 5 Dehydration of ricinoleic acid

Transesterification

Transesterification of vegetable oils refer to the breaking down of vegetable oil molecules by reacting an alcohol with an ester (in the vegetable oil) in which the glycerol functional group from the triglyceride is removed and replaced by an alcohol producing a biodiesel (Scheme 6). The catalysts used in the transesterification are often acid catalysts (e.g. HCl, H2SO4, H3PO4) even though base catalysts (e.g. KOH, NaOH, CH3OK, (CH3O)2Ca, CaO) or heterogeneous catalysts such as zeolites or enzyme catalysts are also used.

Transesterification reactions are reversible and therefore an excess alcohol is usually used to shift the equilibrium to the formation of the biodiesel. Generally transesterification reduces the molecular weight and thus reducing the viscosity of the castor oil which is not required in the biodiesel. Transesterification also increases the volatility while maintaining the cetane number and heating value of the biodiesel. The increased production of biodiesel from vegetable oils has led to the overproduction of glycerol. For instance, Worldwide crude glycerol derived from biodiesel conversion has increased from 200, 000 tonnes in 2004 to 1.224 million tonnes in 2008.

Due to this overproduction of glycerol, scientists are finding new applications for refined and crude glycerol. Examples of such applications include but are not limited to the use of crude glycerol in animal feeds both for ruminants and non ruminant's animals. Feed stock for fermentative production 1, 3-propanediol by Klebsiella pneumonia, biosynthesis of citric acid from crude glycerol by Yarrowia lipolytica ACA-DC 50109 and fermentative conversion of crude glycerol to hydrogen by the bacterium Rhodopseudomonas palustris. Glycerol can also be a good source of various solvents such as propylene glycols, glycerol ethers and esters. Crude glycerol

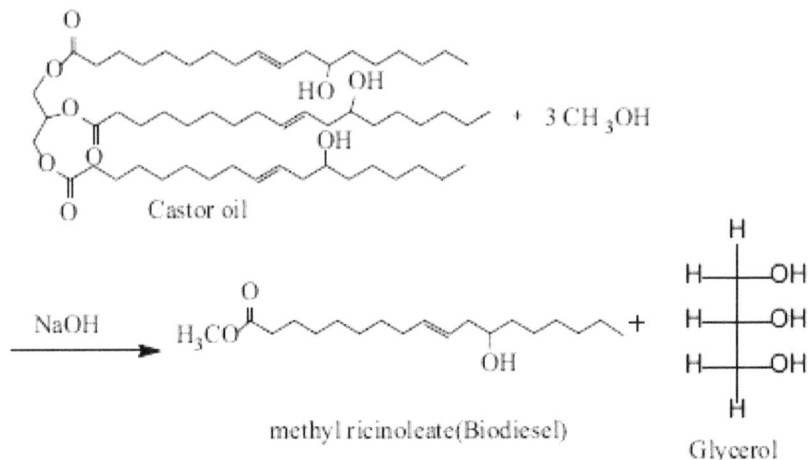

Scheme 6 Transesterification of castor oil

(without any purification) is a green solvent and a reducing agent for metal-catalyzed transfer hydrogenation reactions and nanoparticles formation. Glycerol have also shown potential as a high-boiling-point organic solvent to enhance enzymatic hydrolysis of lignocellulosic biomass during atmospheric autocatalytic organosolvent pre-treatment. Synthesis of aliphatic polyesters from glycerol by reacting it with adipic acid (Scheme 7) is also reported elsewhere.

Sulphation of castor oil

Sulphation refers to the introduction of SO3 group into an organic compound to produce the characteristic C-OSO3 configuration. Sulphation of castor oil produces sulphuric acid esters (Turkey-red oil) in which the hydroxyl group of ricinoleic acid has been esterified (Scheme 8). The reaction is done by treating raw castor oil at room temperature or at temperature less than 308 Kwith concentrated sulphuric acid for 3–4 h. Turkey-red oil is widely used in textile and cosmetics industries by producing synthetic detergents in the recipes/formulations of lubricants, softeners, and dyes. In addition, Turkey-red oil is an active wetting agent in dyeing and in finishing of (Scheme 8) cotton and linen. It is also used in bath oil recipes along with natural or synthetic fragrance or essential oils or in shampoos.

Castor oil based polymers

The depletion of fossil fuels and environmental issues has necessitated researchers to focus their attention and efforts to the utilization of renewable resources as raw materials for the synthesis of polymeric materials. Biobased polymers offer a number of advantages over polymers prepared from petroleum-based monomers as they are cheaper, readily available from renewable natural resources and they possess comparable or better properties.

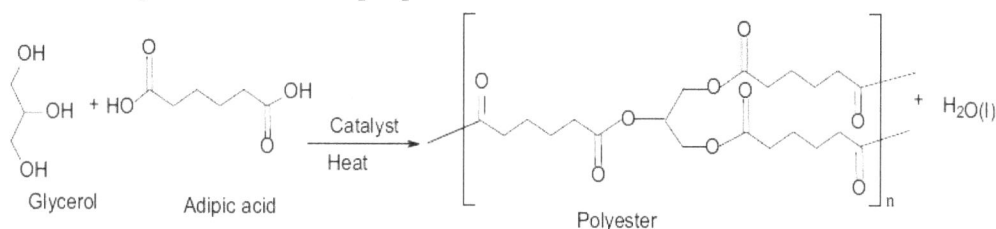

Scheme 7 Synthesis of aliphatic polyester from glycerol and adipic acid

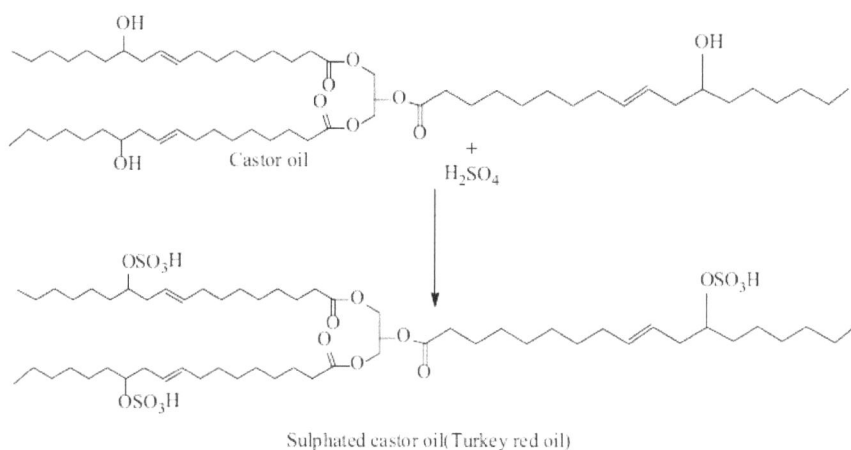

Sulphated castor oil(Turkey red oil)

Scheme 8 Sulphation of castor oil into Turkey red oil

Some bio-based polymers are biodegradable, nontoxic and have low carbon footprints. Polyamides, polyethers, polyesters and interpenetrating polymer networks have been synthesized from castor oil . Most of these castor oil polymers are particularly on the production of polyurethanes, polyamides and polyesters. In another development, the synthesis of interpenetrating polymer networks based on polyol modified castor oil polyurethane and poly(2-hydroxyethylmethacrylate) has been reported.

Ozonolysis of castor oil followed by reduction (Scheme 9) produces triglycerides of 9-carbon fatty acids with terminal hydroxyl groups. The 9-carbon fatty acids can be used as monomers in the preparation of condensation polymers such as polyurethane, polyethers and polyesters.

Polyurethane from castor oil monomers

Polyurethanes (PU) are polymers containing urethane linkages (–NHCOO–) in the main polymer chain. They are among the most important and versatile classes of polymers as they can vary from thermoplastic to thermosetting materials. The industrial production of polyurethanes is normally accomplished through the polyaddition reaction between organic isocyanates and compounds containing active hydroxyl groups, such as polyols. From an environmental viewpoint, this method is not advantageous because it uses highly reactive and toxic isocyanates, which are commonly produced from an even more dangerous component, phosgene. In the search for green routes to the keypolyurethane intermediates, fats and oils offer important alternatives for the production of diols, polyols, and other oxo chemicals, thus, enabling to substitute petrochemicals. Environmentally friendly production of polyurethanes is achieved using plant-derived diols and diisocyanates or using nonisocyanate chemistries. Polyurethanes prepared from vegetable oils exhibit a number of excellent properties that are attributable to its hydrophobicity. Castor oil as a source of polyols, is increasingly finding application in the manufacture of polyurethane.

Polyurethane networks based on castor oil as a renewable resource polyol and poly(ethylene glycol) (PEG) with tunable biodegradation rates for biomedical implants and tissue engineering is documented elsewhere. The synthesis involved the reaction of epoxy-terminated polyurethane prepolymers (EPUs) from castor oil with 1,6-hexamethylene diamine curing agent. This is interesting given that there are a limited number of naturally occurring triglycerides which contain the unreacted hydroxyl groups and castor oil being the only commercially-available natural oil polyol that is produced directly from a plant source as all other natural oil polyols require chemical modification prior to their use.

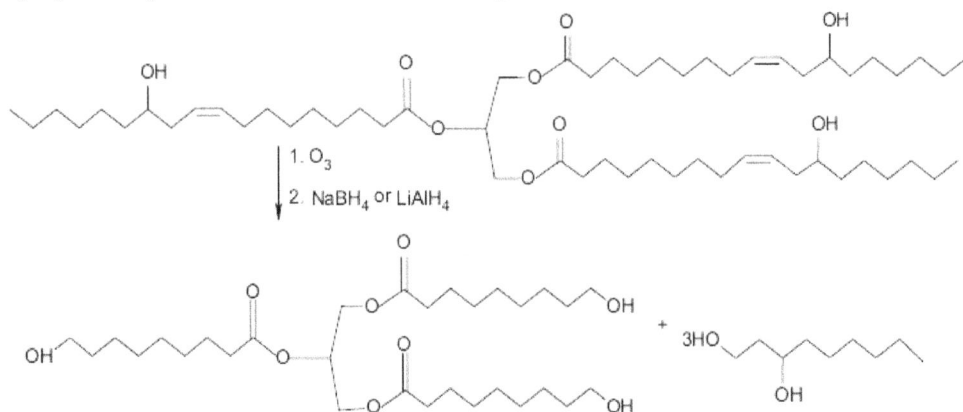

Scheme 9 Ozonolysis and reduction of castor oil

Polyurethane derived from castor oil find their applications in areas such as biomedical implants, coatings, cast elastomers, thermoplastic elastomers, rigid foams, semi-rigid foams, sealants, adhesives and flexible foams.

Methoxycarbonylation of undecylenic acid derived from castor oil

Methoxycarbonylation of plant oils to form diesters is a crucial discovery towards making polymer precursors from plant oils. Pd2(dba)3, 1,2-bis(ditertiarybutyl phosphinomethyl)benzene (DTBPMB), and methane sulfonic acid in methanol have been reported to be an effective catalytic system for making linear diesters from plant oils. Subjecting plant oils to this catalytic system, the double bond located at the center of the molecule is isomerized to the terminal end at ω position with the ester functionality and is trapped by carbonylation to form α-ω diester (Scheme 10). Undecylenic acid with the double bond in the terminal position is a product of pyrolysis of ricinoleic acid from castor oil.

Methoxycarbonylation of undecenoic acid or esters produces dimethyl 1, 12-dodecanedioate, which is a component of Nylon-12, 12. Overall, growth of biopolymers from castor oil industries makes the oil potential for it to play a much larger role in the world economy on polymers and to humanity.

Green synthesis of nanomaterials using castor oil or ricinoleic acid

The high percentage of ricinoleic acid and its structural features makes castor oil capable of forming covalent dative bonds with active surface dangling orbitals of chalcogenides quantum dots. Green synthesis of chalcogenides nanomaterials using castor oil and its isolate ricinoleic acid as eco-friendly bio-based capping agents have recently been reported. This is environmentally interesting because the use of castor oil and ricinoleic acid as both capping and dispersing agents, eliminate the need for the use of air-sensitive, toxic and expensive chemicals such as trioctylphosphine (TOP), trioctylphosphine oxide (TOPO) and alkyl amines. It is worth noting that the boiling points of castor oil and ricinoleic acid are 586 and 685 K, respectively and thus they are simple to work with since they are liquid at room temperature. Literature reports the high ability of castor oil to prevent agglomeration of the synthesized nanoparticles due to the presence of long-chain hydrophobic moieties, thus forming ultra-small, well dispersed and stable quantum dots for a long period of time.

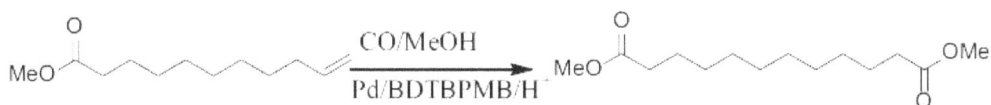

Scheme 10 Synthesis of dimethyl 1, 12-dodecanedioate from methyl undec-10-enoate

Some of these nanomaterials synthesized using castor oil or ricinoleic acid can be suitable for biological and medical applications because no toxic reagents are used in their preparation.

CONCLUSION

The diversity of chemicals and products produced from castor oil has proven that castor is an important and potential non-edible oilseed crop. The great utilitarian value in industry, agriculture, cosmetics and pharmaceutical sectors is a direct proof that castor oil is a potential bio-based starting material. The presence of a hydroxyl group, carboxylate and double bonds in the ricinoleic acid, imparts unique properties for the derivatization of castor oil into vital industrial raw materials.

It has been shown how castor oil can be used as a renewable bio-based raw material for the production a multitude of functional materials. It is equally noted that the diverse possibilities of castor oil transformation mainly depend on the presence of the three functional groups.

This review has further shown that castor oil is a potential alternative to petroleum-based starting materials for the production of wide range of industrial materials. It can also be seen that apart from the oil's unique chemical structure and environmental considerations, the worldwide growth in castor oil demand is due to its easy availability, low cost, non-food competition. It has been observed in the discussion that castor oil is more than just a bio-based raw material in great demand by the chemical industries but its use as a fuel is also seen when transesterification is done. The worldwide increase in the production of castor seeds and castor oil testifies the huge potential as a green bio-resource for chemical transformations because castor oil can be used as the starting material for producing a wide range of end-products.

6 Chapter

EPOXIDIZED SOYBEAN OIL (ESO)/POLY(LACTIC ACID) COMPOSITES

INTRODUCTION

Recently, biodegradable and biobased polymers were extensively used to replace nonrenewable and petroleum- based analogues to reduce plastics waste pollution. Poly (lactic acid) (PLA) is biobased polyester that is originated from starch-rich products including corn and wheat via a fermentation process. PLA possesses attractive properties such as high strength and modulus, easy process ability, and bio - degradability.

Many studies have enabled the production of PLA products with a large-scale level and their various applications in packaging, medicines, textiles, automobiles, etc. However, the intrinsic brittleness of PLA is the main bottleneck that inhibits its wider application.

Blending with other monomers or polymers is easily processed and effective in toughening PLA. Vegetable oils and their derivatives including epoxidized soybean oil (ESO) are attractive materials in blending with PLA due to their flexible molecular chains. However, the toughening efficiency is prevented by the immiscibility of these derivatives with PLA. The interfacial compatibility between soybean oil-based derivatives and PLA was greatly increased by using a soybean oil-PLA star polymer as compatibilizer, which contributed to significantly increased fracture elongation and tensile toughness of soybean oil/PLA blends. A poly(isoprene-b-L-lactide) block copolymer was also used to improve the tensile toughness of polymerized soybean oil/PLA blends. It is reported that dynamic vulcanization is a reactive process that increases the compatibility of each components in PLA-based blends during melt compounding. The polymerization of ESO was initiated with a cationic initiator during the blending of ESO with PLA to produce fully biobased ESO/PLA blends with high toughness. A sustainable sebacic acid-cured ESO phase was introduced into PLA matrix to increase the toughening efficiency of ESO on PLA blends. Similar work was done on PLA matrix incorporated with diisocyanate-crosslinked castor oil phase. In our previous work, tannic acid (TA) was used as a green crosslinker for

ESO during the dynamic vulcanization of PLA with ESO to generate highly tough TA-c-ESO/PLA blends. As proposed, the possible reactions in the TA-c-ESO/PLA blends include 1) the phenolic –OH groups from TA with the epoxy rings from ESO (Reaction I), 2) the formed –OH groups from Reaction I with other ESO epoxides (Reaction II), 3) the terminal –COOH groups from PLA with the –OH groups from TA (Reaction III), and 4) the –COOH/ –OH groups of PLA with the epoxy rings of ESO (Reaction IV). These reactions resulted in a stable TA-cross linked-ESO elastomer formed within PLA matrix as a dispersed phase and an increased interfacial bonding between the TA-c-ESO phase and PLA, which significantly increased the fracture elongation and tensile toughness of the PLA blends, while the tensile strength and modulus were dramatically reduced.

Nanofillers, such as cellulose nanocrystals, graphene, montmorillonite, TiO2 nanoparticles, halloysite nanotubes, and carbon nanotubes (CNTs), have been incorporated into the PLA matrix for the preparation of PLA-based nanocomposites with significantly improved mechanical strength. CNT/PLA nanocomposites are particularly attractive due to the ability of CNTs to induce good mechanical strength and stiffness as well as superior thermal and electrical properties.

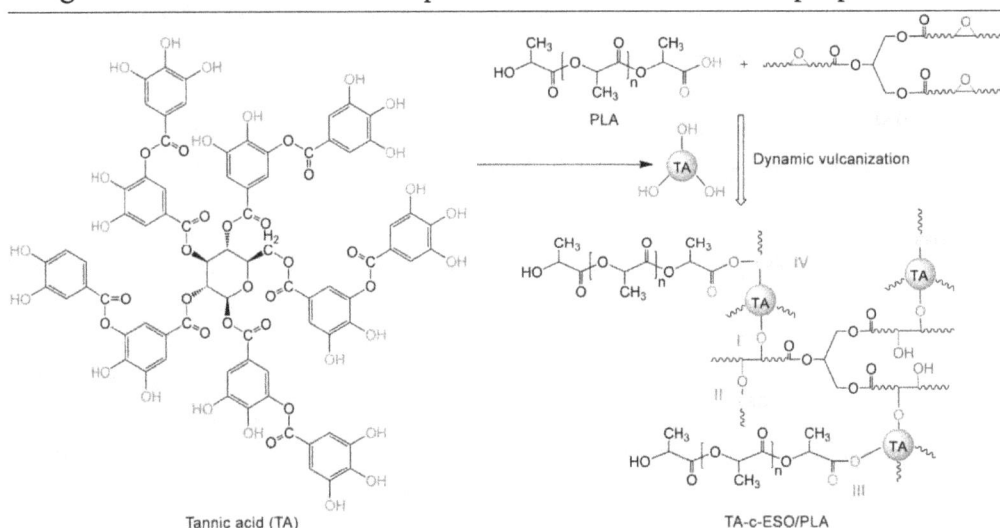

Figure 1. Proposed reactions among TA, ESO and PLA during dynamic vulcanization

Our previous study found that the PLA nanocomposite with 3 wt% CNTs obtained from twin-screw extrusion achieved significantly improved tensile strength and modulus as well as electrical conductivity. The key challenges for preparing CNT/PLA nanocomposites are the poor dispersion of CNTs in PLA matrix and filler-matrix interfacial adhesion, which could be alleviated by functionalization of CNTs. However, incorporation of CNTs dramatically deteriorated the impact strength and fracture elongation of the resulting PLA nanocomposites. Investigations on the toughening of CNT/PLA nano - composites have been addressed by blending with plasticizers, elastomers, and other flexible polymers. For example, novel electroactive shape memory polymer nanocomposites were prepared from CNTs and ESO or epoxidized linseed oil plasticized PLA . More importantly, the addition of CNTs into an immiscible polymer blends is effective in improving the fracture toughness of the blends and the toughening efficiency depends on the selective distribution of CNTs in the blend components. The toughness of PLA blends was greatly improved by incorporation of pristine and functionalized CNTs, along with the addition of different toughening phases, such as thermoplastic polyurethane (TPU), polycarbonate (PC), poly(butyleneadipate- co-terephthalate), poly(3-hydroxybutyrateco-4-hydroxybutyrate), polystyrene, poly(ethylene-co-vinyl acetate), and natural rubber. The mechanical properties and miscibility of PC/PLA blends were greatly increased by simultaneous addition of a compatibilizer and CNTs.

Highly conductive PLA/TPU/CNTs nanocomposites with a low percolation threshold and excellent stiffness- toughness balance were obtained via inducing the formation of co-continuous structure in the PLA/ TPU blend matrix.

To further develop ESO/PLA-based nanocomposites with superior strength-toughness balance, TA and CNTs were simultaneously incorporated into the composite systems. The TA was used as a cross linker to induce the cross linking of ESO and form highly tough TA-c-ESO/PLA blend, while the CNTs was expected to impart the obtained CNT/TA-c-ESO/PLA nanocomposites with good mechanical strength and modulus as well as electrical properties. Fixed TA usage (10 wt% TA-c-ESO with a molar ratio of –OH to epoxy of 0.6) and various amounts of CNTs (0.5– 10 wt%) were added into the ESO/PLA products to investigate the influence of CNTs content on the performance of the resulting CNT/TA-c-ESO/PLA nano -composites. The synergistic effects of CNTs and TAc- ESO on the reinforcing and toughening of the nanocomposites were also investigated.

MATERIALS AND METHODS

Materials

PLA (Ingeo 3001D) was given by Nature Works Japan. ESO (epoxy value: >6) was purchased from Aladdin Industrial Corporation (China). Multiwalled CNTs (NANOCYLR NC7000™) with an average diameter of 9.5 nm and an average length of 1.5 μm were obtained from Nanocyl Japan. TA was purchased from Nacalai Tesque, INC (Japan).

Preparation of CNT/TA-c-ESO/PLA nanocomposites

The preparation of TA-c-ESO/PLA blends was described in previous work. Briefly, TA-dispersed ESO solution was prepared by dissolving TA (2.92 g) and ESO (17.08

g) (a molar ratio of –OH groups to epoxy groups of 0.6) in acetone (20 g). Then, the TA-dispersed ESO solution was mixed with the oven-dried PLA pellets (180 g) at a weight ratio of 1:9 on a twin-screw extruder (Technovel Corporation Japan) to perform dynamic vulcanization process.

The extruded TA-c-ESO/PLA blends were cooled in water and oven-dried for composite preparation. The obtained TA-c-ESO/PLA blends were mixed with CNTs at different weight ratios of 0.5, 1, 3, 5, 7, and 10 wt%, respectively, to fabricate CNT/TAc- ESO/PLA nanocomposites. The nanocomposites were abbreviated as xCNT/TA-c-ESO/PLA, where the x means the weight fraction of CNTs in the nano - composites. The mixing process was performed on the same twin-screw extruder as the procedure of TA-c-ESO/PLA blend preparation. Composite specimens were obtained from an injection molding machine (NP7-1F, Nissei Plastic Industrial, Japan). The nozzle temperature and mold temperature were 170 and 40 °C, respectively. Further, specimens from TAc- ESO/PLA blend and PLA nanocomposite with 5 wt% CNTs (5CNT/PLA) were prepared under same conditions for comparison.

CONCLUSION

High biobased content, strength-toughness balanced, and electrically conductive CNT/TA-c-ESO/PLA nanocomposites were fabricated by dynamic vulcanization of PLA with ESO by using TA as a cross linker and CNTs as reinforcement. Results indicated that the obtained TA-c-ESO/PLA blend presented a typical sea-island structure and the incorporated CNTs were mainly dispersed in the continuous PLA phase. The combination of TA-c-ESO phase and CNTs resulted in an improved crystalline of PLA phase due to their heterogeneous nucleation effects. The tensile strength and modulus, storage modulus, and thermal stability of the nanocomposites were increased with the introduction of CNTs; however, the added CNTs led to significant decreases in fracture elongation, tensile toughness, and damping ability of the nanocomposites. This indicated that the toughening effect of TA-c-ESO phase was reduced by the addition of stiff CNTs. The formed sea-island structure in TA-c- ESO/PLA blends contributed to improving the electrical conductivity of the nanocomposites. The PLA nanocomposites having 10 wt% TA-c-ESO phase and 5 wt% CNTs achieved the optimum strength toughness balance and favorable electrical conductivity.

In a word, a facile and cost-effective method to toughen and strengthen PLA products was provided by using TA to induce cross linking of ESO and CNTs as reinforcing fillers.

7
Chapter

SOYBEAN OILS FOR USE AS BIOBASED THERMOSET RESINS

INTRODUCTION

In search of alternative raw materials for resins and polymers, the plant oils have been identified as versatile candidates. Plant oils, or natural triglyceride oils, are abundant in most parts of the world, and they are ideal alternatives to crude oil based feed stocks. Triglycerides are esters of fatty acids and glycerol, and the fatty acid chain contains typically 14 to 22 carbons in length of which 0 to 6 are unsaturated. There are many plants that can be used for oil production.

Soybean, rapeseed, sunflower, linseed and sunflower are some examples. The exact chemical composition varies a lot between the plants, and there are also individual variations among the plants. The plant oils main use is in the food industry, but there is also a traditional use as binders for coatings and inks, lubricants as well as plasticizers. As such the plant oils cannot be used in rigid and structural plastic products, due to their low molecular weight. This is however possible, if the plant oil triglyceride is functionalized so it can undergo further chemical reactions. There are several chemical reactions available, which can be utilized for the functionalisation into a thermosetting resin. Maleinisation, acrylation, hydroxylation and anhydrisation are some reported methods. The obtained resin can then be cross-linked by various methods, and a rigid plastic material is then obtained.

In this paper we will describe synthetic modification of epoxidised soybean oil by reaction first with methacrylic acid, and then with methacrylic anhydride or acetic anhydride. The reactions were characterized by FT-IR spectroscopy, 1H and 13C NMR spectroscopy. The curing of the resin was investigated by differential scanning calorimetry (DSC), after adding a free radical initiator.

EXPERIMENTAL

Materials.

Epoxidised soybean oil (EDENOL D81) was supplied by Cognis GmbH, Germany. Methacrylic acid, (99 %), methacrylic anhydride (94 %) and acetic anhydride (99 %) were used for the modification. Nmethylimidiazol (99 %) (GC) was used as a catalyst, and hydroquinone (99 %) was used as cross-linking inhibitor. All chemicals were purchased from Aldrich Chemical Company, USA. Chloroform (99 %) was supplied by Fisher Scientific, Sweden. tert-Butyl peroxy benzoate was used as the free radical initiator for the curing experiments and it was supplied by Adonox PB, Sweden.

INSTRUMENTATION.

The epoxidised soybean oil and the synthesized resins were characterized and analyzed by FT-IR and 1H respective 13C NMR spectroscopy. FT-IR analysis was carried out on Nicolet spectrometer (1Mw HeNe Laser 633nm), the DSC scanning was done using (DSC Q-200 seriesTM model) while the NMR analysis was performed by dissolving the samples in chloroform and run them on a 400 MHz instrument.

Synthesis of methacrylate modified soybean oil.

Epoxidised soybean oil (13.2 g, 0.01 mol) was heated from room temperature while methacrylic acid (4.3 g, 0.05 mol) containing hydroquinone (0.01 g, 0.25 wt %) was added during 30 minutes. The reaction mixture was heated under reflux for 8 hours at 120°C while being constantly stirred. Excess methacrylic acid (1.7 g, 0.02 mol) containing hydroquinone (0.004 g) was then added during 5 minutes and the reaction was allowed to proceed for another 4 hours at the same temperature. The mixture was then cooled to room temperature and purified by extraction in chloroform and further washed with 5 % anhydrous sodium carbonate. The organic layer was dried over anhydrous sodium sulphate and evaporated using rotational distillation. The obtained product, methacrylated soybean oil (MSO) was then isolated.

The product weighed 14.9 g and was further purified by gel filtration. The initial synthesis was up-scaled, first 4 fold, and then 12 fold twice. The total amount of obtained MSO was about 400g.

Further modification with methacrylic anhydride.

15.7 g of MSO was dissolved in chloroform and heated under reflux condenser in a 3 neck round bottomed-flask at 55°C. Methacrylic anhydride (7.7 g, 0.05 mol) was added in drop wise during 15 minutes, after which the temperature rose a little above 60°C, the boiling point of chloroform. Nmethylimidazol (1 wt %, 0.25 g) was used as a catalyst. The reaction proceeded for 3 hours being constantly stirred. The reaction mixture was allowed to cool to room temperature and the extraction was done as for step 1. The product MMSO obtained weighed 16.0 g. The synthesis was then upscaled 4 fold, and totally 55 g product was obtained.

Further modification with acetic anhydride

10.2 g MSO was dissolved in chloroform and heated under reflux in the presence of N-methylimidazol (1 wt %, 0.25 g), acetic anhydride (5.1 g, 0.05 mol) was then added drop wise. The synthesis proceeded as previously described. 10.6 g of product AMSO was obtained. The synthesis was then upscaled 4 fold and this gave 55g of the wanted product.

Purification of products.

Gel filtration of the various products of the syntheses was done using silica gel 60. The products were dissolved differently in chloroform and filtered through the column. Isolation of the various products was done using rotational distillation and the purified products were analyzed.

Vicosity measurements

The viscosity of the neat uncured resin was determined using a Bohlin rheometer CS30 from Malvern Instruments Ltd. All measurements were done with a cone plate configuration with a truncated cone (Ø15 mm, 5.4°). The characterisation was done by stress viscometry. For each temperature, three different levels of the stress were chosen and the viscosity was calculated.

Curing experiments

The resins were mixed with t-Butyl peroxybensoate as free radical initiator and placed in an oven heated to 160°C (thermal curing). Samples were withdrawn at 5 min, 10 min, 1 hour, 2 hours, 4 hours, 19 hours and 24 hours and analyzed by DSC and FT-IR. The data from the first dynamic DSC scans were analyzed for possible exothermic peaks. For the photocuring, the resins were blended with the photoinitiator, Irgacure 819, and exposed by blue light for 0s, 6s and 90s, using a Heraflash light curing unit. The remaining heat exotherm was analysed by dynamic DSC scans.

CHARACTERISATION

The characterisations of these thermoset resins were done using FT-IR and NMR spectroscopy. By using these methods the obtained structures could be characterized and verified. Thermal and photocuring were done on the obtained resins and the cross linked resins were also characterized by FT-IR spectroscopy.

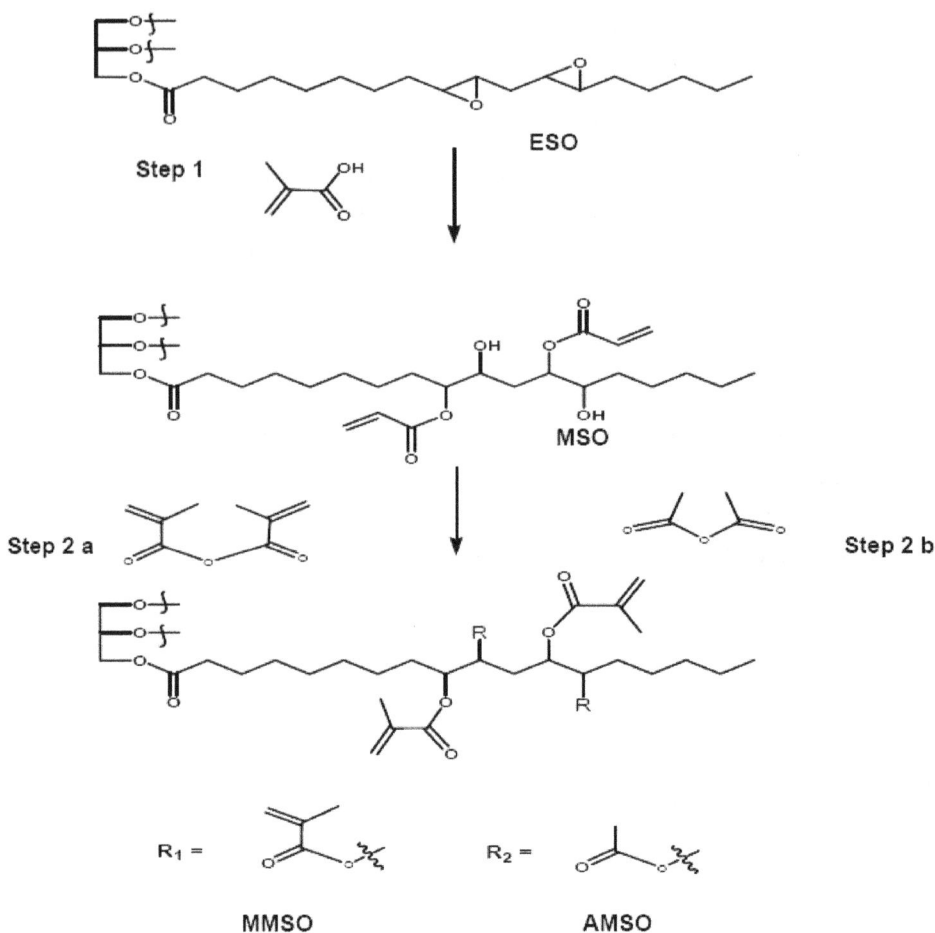

Figure 1. Reaction scheme for the modification of epoxidised soybean oil

RESULTS AND DISCUSSIONS

Fourier-transform IR spectroscopy.

Table 1 gives the summary of the functional groups identified from the FT-IR spectra of all the samples available for analysis.

Table 1. Summary of the FT-IR spectra

Wave Number (Cm⁻¹)	Functional Group Assigment	Samples
3491	-OH	MSO
1740	C=O	ESO-MSO
823 and 841	Epoxy Group	ESO
1200 and 1275	Acetylgroup	AMSO

The absorption bands of the epoxy group at 823 and 841 cm-1 present in ESO disappear in the IR spectra of MSO, while double bands appear at 1739 and 1716 cm-1. The hydroxyl absorption was around 3491 cm-1 and 3472 cm-1 in the spectra of MSO respectively, while this was absent in the spectrum of ESO. The absorption at about 1630 cm-1 which is characteristic of carbon double bond can be seen in the spectra of MSO, while this is absent in the spectrum of ESO. The IR spectra of the resins MMSO and AMSO, show that the hydroxyl peaks disappear and the absorptions at about 1715 and 1730 cm-1 characteristic of carbonyl group and the carbon double bond also at 1636 cm-1 are visible, which indicates a successful reaction. The IR spectrum of AMSO shows absorptions at 1217 and 1235 cm-1 which could be due to acetyl functional group.

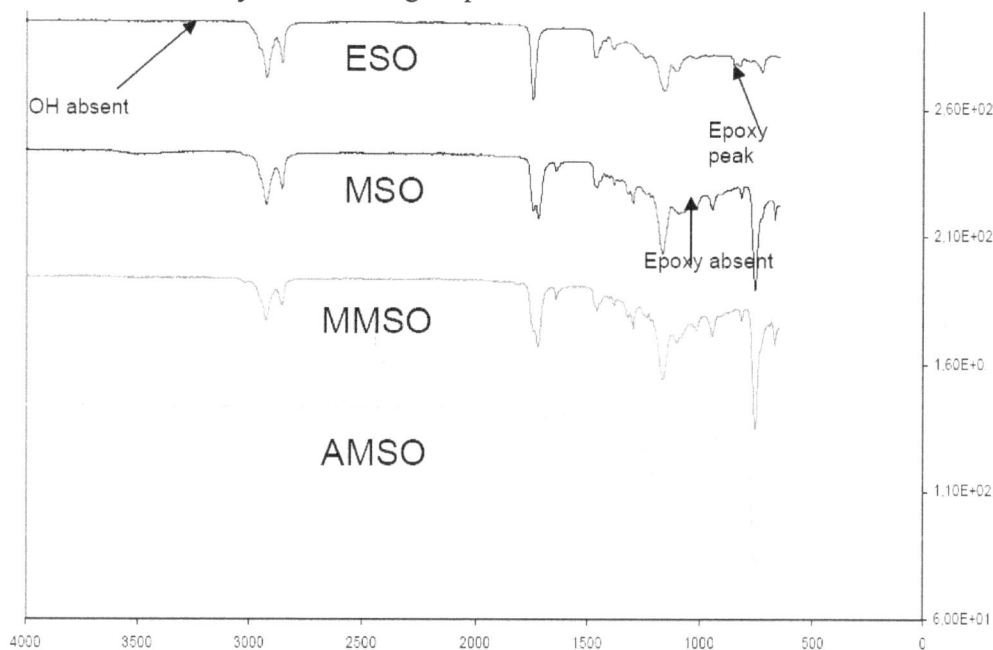

Figure 2. FT-IR Spectra comparison of ESO, MSO, MMSO and AMSO

NMR SPECTROSCOPY

The NMR analysis verified that the wanted reactions occurred. The epoxy carbons in the ESO sample are in the 52 - 57 ppm region. These signals are missing in the AMSO sample and several new peaks around the 72-85 ppm area can be seen. These are presumed to be due to the following reaction:

The 13C NMR spectra of ESO and AMSO were also used to confirm the methyl group assignments and also to determine the ratio of acetate and methacrylate methyl groups (and therefore mole ratio of acetate and methacrylate groups) in the AMSO sample. The ratio of acetate and methacrylate methyl groups in the AMSO sample was found to be 1:1.

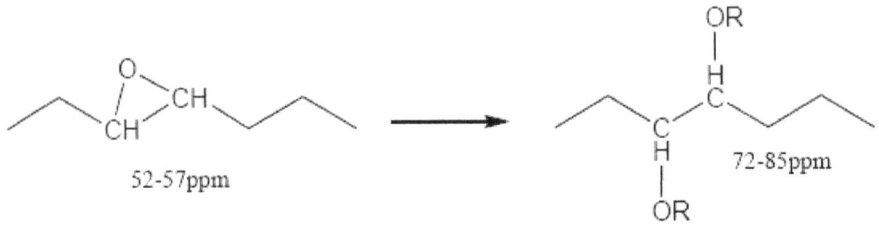

52-57ppm   →   72-85ppm

R = H or acetate or methacrylate

This was also confirmed by 1H NMR spectrum of the ESO. The epoxy groups give signals around 3 ppm, and in the modified resin samples these have disappeared, and subsequently new signals are formed around 4.0 and 5.0 ppm. The peaks around 4.0 ppm are probably due to CH-OH groups, while the ones around 5.0 ppm are due to the CH-OR groups (R = acetate or methacrylate). The degree of epoxy group conversion determined from the 1H NMR data is given in Table 2.

Table 2. Loss of epoxy protons from the samples

Sample	Epoxy H/molecule	Loss of epoxy H's	Residual epoxy H's
ESO	7.95	–	100%
MSO	0.46	94.2%	5.8%
MMSO	0.43	94.6%	5.4%
AMSO	0.64	92.0%	8.0%

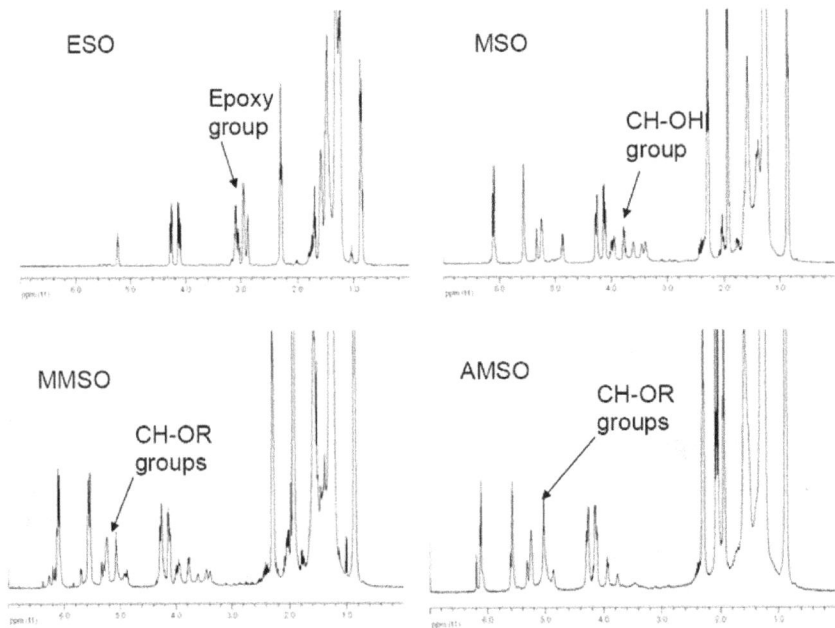

Figure 3. 1H- NMR Spectra comparison of ESO, MSO, MMSO and AMSO

CURING RESULTS

The DSC scan for the sample withdrawn up to 10 min showed an exotherm peak indicating incomplete cure. Longer curing times gave no exothermic peaks, meaning that the curing was complete. The FT-IR spectra of the samples drawn at 5 and 10 min showed carbon-carbon double bond peak at about 1635 cm-1 which indicates that the samples are not fully cured, while no carbon-carbon double bonds could be seen in the samples drawn at intervals between 1 hour and 24 hours.

The applicability of the three resins for photocuring was examined. The neat resins were blended with a photo initiator and the resins were exposed to blue light. All the resins can be cured to completion by photocuring. The MSO resin seems to be the most reactive and goes to almost completion with 6 seconds of exposure. The MMSO resin also seems to be feasible for photocuring. The AMSO resin only shows about half of the heat exotherm in comparison to the MSO resin. Considering that both the resins contain one carbon-carbon double bond one would expect both resin to have a similar heat exotherm after o seconds exposure. Besides, considering that the MMSO resin contains two double bonds in contrast to the other two resins, one would expect a higher heat exotherm for this resin as well. It would take more tests to completely characterize these resins in terms of photocuring.

Table 3. Dynamic DSC scans after different exposures.

Resin + 4% Irgacure 819	0 s exposure (J/g)	6 s exposure (J/g)	(90 s exposure)
MSO	36.77	2.1	0
MMSO	34.57	8.00	0
AMSO	16.2	0	0

VISCOSITY

The viscosity of a thermoset resin is very important since it will determine how the resin can be processed. The result of the viscosity measurements are shown in Figure 5. Both the MMSO resin and the MSO resin have low viscosities at room temperature. Especially the MSO resin which have a viscosity of about 0.22 Pas at 25°C. Commercial petroleum based resins, such as unsaturated polyester resins are invariably diluted with styrene in order to lower the viscosity. These tests indicate however that both the MMSO and the MSO resin can probably be processed with infusion techniques such as resin transfer moulding with only a small addition styrene or eventually without styrene.

The viscosity curve of the AMSO resin does not look very reliable. The rheological behavior of this resin needs to be further investigated.

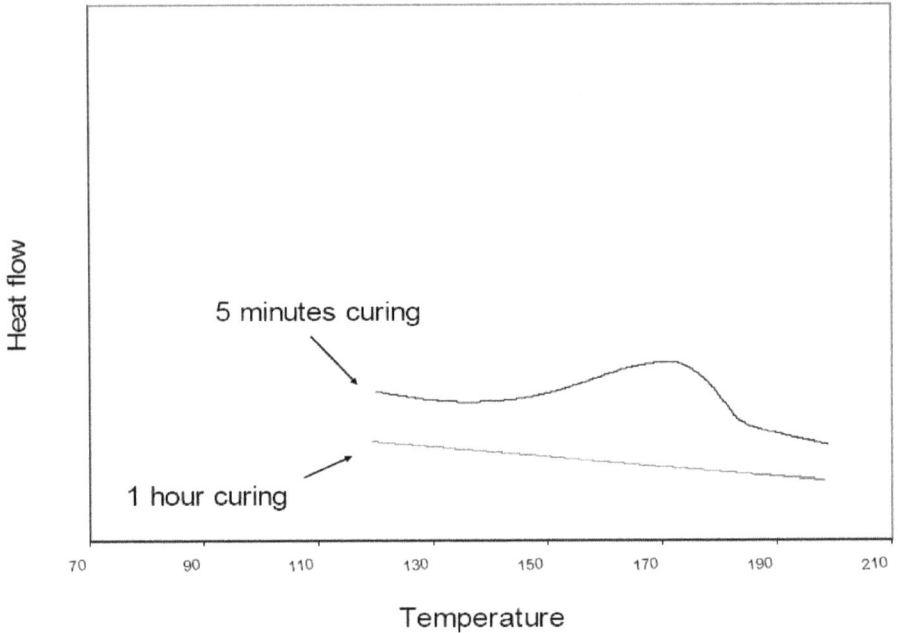

Figure 4. Cured with 2 wt-% tert-Butylperoxybenzoate at 160°C

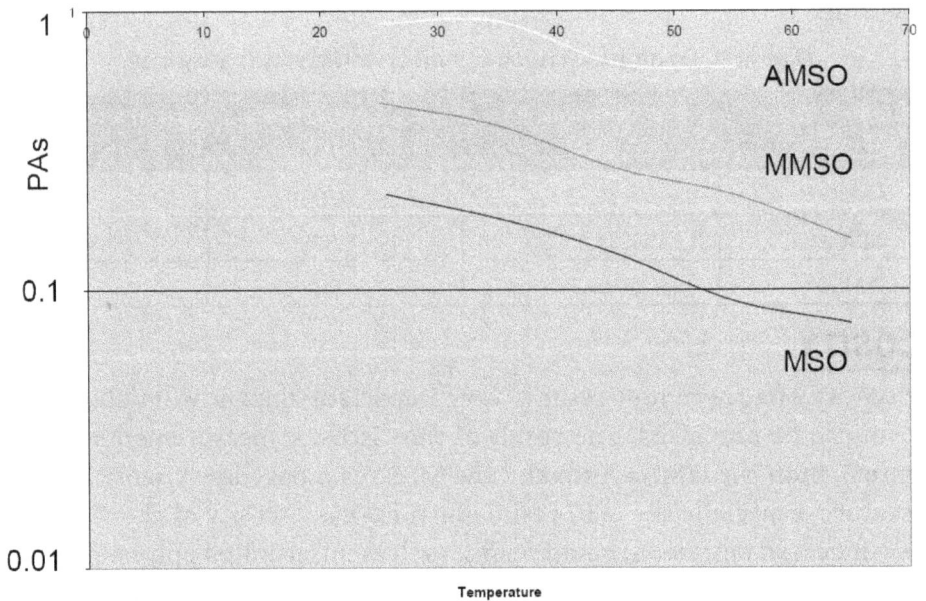

Figure 5. Viscosity measurements

CONCLUSION

Bio-based resins for use in thermoset composites were synthesized from soybean oil. The epoxidised soybean oil was functionalized with methacrylic acid (methacrylation) and the resulted product was further modified using acid anhydrides, the aim was not only to increase the crosslink sites per triglyceride but also reduces the number of free hydroxyl groups, which have a viscosity increasing effect. The methacrylation of the epoxidised soybean oil using a 0.2 molar excess of methacrylic acid could maximize the conversion of epoxide groups to more than 90 %. The further modification with methacrylic anhydride proved to be more effective than with acetic anhydride. A higher conversion of epoxy groups was achieved up to 95 % in the MMSO sample, this was confirmed by the 1H NMR and the ratio of acetate and methacrylate methyl groups in the AMSO sample was found to be 1:1, this was confirmed by the ^{13}C-NMR. The methacrylated epoxidised soybean oil MSO was mixed with 2 wt % tert-butylperoxybenzoate and was cured thermally at 160°C for 24 hours, however a fully cross linked thermoset polymer was achieved after 1 hour, and this was confirmed by the DSC. No structural changes observed in the cross linked polymer inspite of the prolonged cure in the oven, this was also confirmed by the FT -IR.

8
Chapter

BIO-RENEWABLE POLYMER COMPOSITES FROM TALL OIL-BASED POLYAMIDE AND LIGNIN-CELLULOSE FIBER

INTRODUCTION

Polymers and composites derived from biorenewable resources have received extensive attention as sustainable alternatives to petroleum-based polymers due to the increasing cost of fossil fuels and various environmental concerns. It was estimated that the petroleum resources will be depleted within one hundred years. The traditional petroleum-based polymers are not biorenewable, and most of them are not biodegradable. These widely used petroleum-based polymers have introduced many environmental problems, such as the emission of greenhouse gases and white pollution.

Many thermoplastics and thermosets based on bio-renewable resources have been developed. Polylactide (PLA) is a widely used thermoplastic produced from the fermentation of corn and sugar feedstocks. Because PLA is biorenewable and biodegradable, it has been used in packaging and biomedical applications.

On the other hand, poly hydroxylalkanoates (PHA), is a class of polyesters produced from bacteria. Bacteria produce PHA for carbon and energy storage. PHA is biorenewable and biodegradable, and it also possesses high biocompatibility. PHA has been used as drug carriers and scaffold materials in tissue engineering. PHA possesses mechanical properties similar to those of polypropylene; however, the high cost and the brittleness of PHA have limited its applications as a general plastic.

Plant oils, such as soybean oil, castor oil, and tung oil, are very popular starting materials for synthesizing biorenewable thermoplastics and thermosets, since they carry many chemical reactive sites such as double bonds, hydroxyl groups, epoxide groups and ester linkages. Polyamide is a class of polymers that has been widely used in textiles, automotive, electrical, and adhesive application. Some of the most seen polyamides are Nylon 6 and Nylon 6, 6, which are produced from petroleum-based chemicals.

Polyamides based on vegetable oils have also been synthesized. Polyamide-11, a castor oil-based polyamide produced by Arkema, can be synthesized via polycondensation of 11-aminoundecanoic acid (a fatty acid derived from castor oil). To synthesis 11- aminoundecanoic acid, castor oil is saponified under basic condition and neutralized to produce ricinoleic acid. The ricinoleic acid is then esterified using methanol and ethanol to produce ricinoleic ester. The ester of ricinoleic acid is then heated to 500°C to yield undecylenic acid. The undecylenic acid is then brominated using HBr with the presence of peroxide catalyst, followed by amination reaction using ammonia to yield 11- aminoundecanoic acid – the building block of Polyamide 11. Another way to synthesis biorenewable polyamide involves the use of vegetable oil-based dimer. Plant oils are first saponified into fatty acids and then converted to dimer acid. The dimer acid contains two carboxylic acid functional groups. Polyamides can be produced after adding diamine to react with the carboxylic acid group from the dimer acid. Hablot et al. synthesized polyamide based on rapeseed oil dimer acid and 1, 2-diaminoehtane, 1,6- diaminohexane or 1,8-diaminooctaine.

The obtained polyamides are semi-crystalline polymers with a degree of crystallinity around 10%. The melting points of the resulting polyamide ranges from 79°C to 105°C, and the glass transition temperature was in the range of -17°C to -5°C . This polyamide is soft and flexible with more than 300% maximum tensile strain. Fan et al. prepared a series of polyamide based on soy-based dimer acids with a glass transition temperature as high as 63°C and a modulus value above 2000 MPa. Moreover, polyamides from tung oil and soybean oil have also been synthesized for the paint industry due to their thixotropic rheological properties.

Biorenewable polymers are generally inferior to petroleum-based polymer in terms of cost and mechanical properties. For example, PLA and PHA are much more expensive than common petroleum-based polymers such as polyethylene and polystyrene, and they are notorious for their inherent brittleness. A common strategy to compensate the drawbacks of biorenewable polymer is by blending them with fillers or fibers to decrease the overall cost and/or to modify the mechanical properties. There are numerous studies about adding fillers or fibers to biorenewable thermoplastics. Plant fibers such as kenaf fibers, jute fibers, and bamboo fibers, and synthetic fibers such as glass fibers and carbon fibers have been added to PLA. Generally, after adding rigid fibers into the PLA matrix, the strength and modulus will both increase if strong interfacial adhesion can be achieved. The effects of organic fillers such as flours, starches, rice straw, lignin, and cellulose on the properties of biorenewable thermoplastics have been extensively studied.

Moreover, adding agriculture based fillers can also increase the biodegradation rate of the biodegradable polymers. For example, adding distiller dried grains (DDGS) – a cereal co-product of the corn-ethanol industry into PHA can not only decrease the overall cost of the composites, but also increase the biodegradation rate significantly.

There are three major components in the biomass or the cell wall of the plant: lignin, cellulose, and hemicellulose. Lignin is the second most abundant natural resource next to cellulose, and it is a byproduct of the paper and pulp industries and the bio-ethanol industries. Lignin is an amorphous low molecular weight polymer produced from dehydrogenative polymerization of three types of phenols: p-coumaryl, coniferyl and sinapyl alcohols. Lignin makes up about 10 - 30% in wood. In woody plants, lignin offers protection against water, pathogens, pests and enzymatic degradation. Lignin also acts as a binder that holds hemicellulose and cellulose together, providing stiffness to a plant. Lignin has been used as fillers for many thermoplastics. The incorporation of lignin can alter the mechanical properties, thermal stability, and crystallization behavior of the thermoplastics.

On the other hand, cellulose covers about 45 wt% of dry wood, and it is the most abundant natural material on earth. Cellulose is a polysaccharide consisting of D-glucose linked by β-1, 4 linkages. It forms the primary structure component in a plant. Cellulose is a hydrophilic, biodegradable, and semi-crystalline polymer. Hemicellulose is a polysaccharide that makes up 25-30% of wood. It is an amorphous low molecular weight polymer (with molecular weight less than that of cellulose), and it acts as a compatibilizer between cellulose and lignin. Blending plant-based fillers such as cellulose and lignin with biorenewable plastics can not only produce composites with lower cost, but also increase the biorenewable content and strength of the composites.

The objective of this work is to study the thermal, mechanical, and rheological properties, as well as the morphology of a tall-oil based polyamide reinforced with LCF. The fracture surface morphology was studied using scanning electron microscope (SEM). Dynamic mechanical analysis (DMA) and differential scanning calorimetry (DSC) were performed to study the thermo mechanical properties of the PA/LCF composite.

Thermo gravimetric analysis (TGA) was used to investigate the effect of LCF on the thermal stability of the polyamide. In addition, the rheological behavior of the composites was studied using a rheometer. Standard dog-bone shaped specimens for tensile testing were prepared to investigate the change in yield strength, modulus, and elongation after incorporating LCF into the tall oil-based polyamide.

EXPERIMENTAL PROCEDURE

Materials

The polyamide (PA) used in this study was UNI-REZ 2651 supplied by Arizona Chemicals (USA). This polyamide is produced based on tall oil-fatty acid dimer. This PA possesses high flexibility with a maximum elongation above 500%. This PA has a softening point around 95 - 105°C and an amine value of 5. The lignin-cellulose fiber (LCF) was obtained from New Polymer System (New Canaan, CT) in 100-mesh powder form. This filler contains both cellulose and lignin that are extracted from pine trees, and the hemicellulose in the tree is removed by a thermo chemical process.

Composite Preparation

Before compounding, PA and LCF were dried at 60°C for 24 h to remove all the moisture. All the composites were prepared by compounding PA with different compositions of LCF using a twin-screw micro-compounder (DACA Instrument, Santa Barbara, CA). The materials were compounded using a rotational speed of 100 rpm for 10 min. The temperature of the barrel was set to 140°C. The neat PA polymer was also processed using the same conditions so that the samples have the same thermal history.

Composites with the following LCF content were prepared: 10 wt%, 20 wt%, 30 wt%. The nomenclature of the composites is presented as follows: PA-20% represents the composites containing 20 wt% of LCF and 80 wt% of PA. The extruded blends were then compression molded using a Carver Model 4394 hydraulic press (Wabash, IN, USA) to form DMA and tensile testing specimens. The temperature and force for compression molding were set to 140°C and 2 tons, respectively. The specimens were cooled to room temperature under pressure before being taken out of the mold.

Morphological characterization

The extruded blends were cryogenically fractured and viewed under an FEI Quanta FEG 250 scanning electron microscopy to examine to morphology the composites. Before the samples were put into the SEM, all samples were sputter coated with a 5nm-thick layer of iridium. The SEM images were taken at a working voltage of 10 kV under high vacuum.

CONCLUSION

Tall oil-based polyamide was blended with lignin-cellulose fiber (a type of fiber that contains both cellulose and lignin) to produced biocomposites. The effects of LCF on the thermal, mechanical, and rheological properties have been investigated. DMA results indicated that the glass transition temperatures were only slightly affected by the addition of LCF. DMA results also showed that the storage modulus at room temperature increased with increasing amount of LCF.

The enhancement in storage modulus indicated that the reinforcement effect of the LCF fiber. TGA test was performed to study the thermal degradation behavior, and the TGA data showed that LCF decrease the thermal stability of the composites at the 250°C - 400°C range. The dynamic viscosity and shear modulus increased significantly with the increasing LCF content. Tensile testing was used to investigate the mechanical properties of the PA/LCF composites. The Young's modulus and yield strength increased with the additional of LCF, but the strain at break was reduced. To conclude, this chapter demonstrates that the lignin-cellulose fiber can be blended with tall oil-based polyamides via melt processing to produce biorenewable composites with lower cost, higher mechanical properties, and higher biorenewable content.

9

Chapter

LINSEED OIL-BASED THERMOSETS

INTRODUCTION

In a context of sustainable development, the depletion of petroleum resources and the use of hazardous products have led to consider new synthetic pathways toward more eco-friendly toxic polymers. As renewable resources, vegetable oils appear as interesting alternative building blocks. Their strong interest lies in their wide availability, low cost and low toxicity towards human and environment. The nature of the vegetable oils as well as the cultivation conditions strongly affects the structure of triglycerides. A large panel of structures is thus available with a various number of unsaturations, presence of dangling hydroxyl groups and even epoxy functions. Triglycerides contained in vegetable oils are easily functionalized via their double bonds to synthetize various monomers of high functionality which directly gives access to bio-based cross-linked polymers. Moreover, triglyceride monomers with epoxy, hydroxyl, cyclocarbonate, and amine, siloxane, or acrylate groups can be obtained. Epoxidized triglyceride is synthetized through enzymatic process or organic pathway using H2O2/HCOOH and is commercially available. The Michael addition between a nucleophile and an acrylate has drawn much attention due to several advantages. Indeed, this reaction requires mild conditions, high degrees of conversion are easily reached and a large number of commercial monomers are available. This reaction have been widely used to synthetize polymers of varied structures which found utilization in a wide range of applications such as adhesives, coatings, additives, chemical synthesis or bio-medical applications.

Different functional groups are used as donors such as alcohol, amines and thiols. Highly functional monomers are however necessary to afford high performance thermosets. To this extent, vegetable oils are thus interesting candidates. Acrylated epoxidized oils (AEO) are easily obtained by ring opening reaction of epoxides with acrylic acid. This reaction leads to acrylate functions with a vicinal hydroxyl group on the aliphatic chain. AEOs have been widely used via free radical polymerization to synthetize thermosets. Various free radical initiator systems have been explored such as thermal or UV, initiators along with kinetics studies of curing. Thermosetting foams have also been synthetized from acrylated epoxidized soybean oil (AESO) and carbon dioxide using peroxide as free radical initiator. Furthermore, cross-linking of AEOs was also be performed by Michael reaction. Wang et al. especially studied the thia-Michael reaction to synthetize UV-curable thermosetting AESO and studied the effect of different thiols on thermoset properties. Kasetaite et al. synthetized acrylate-thiol thermosets by addition of various thiols on AESO by thermal activation. They compared thermal and mechanical properties of polymers cured with thiols with various functionalities from 2 to 6. The addition of amine to hyperbranched sucrose-soyate via aza-Michael reaction have been performed by Webster's group. They especially studied the effect of the solvent, the ratio, the use of catalysts and different cross-linkers on coatings properties. However, no bulk thermosets have yet been reported via aza-Michael reaction on AEO. Furthermore, hydroxyl groups are known to improve reactivity by intermolecular activation via hydrogen bonds. The use of such activated structure can improve reactivity and allow performing polymerization in mild conditions in order to avoid any catalyst or high temperature. This autocatalytic and sustainable monomer is thus very attractive in terms of green chemistry.

Therefore the present paper reports the study of reaction between amine and acrylated epoxidized linseed oil (AELO) without neither solvent nor catalysts. A first study was performed on model molecules to highlight the strong reactivity of AELO with amines and to identify the role of hydroxyl groups. Amines with various structures were then chosen to study reactivity with AELO by differential scanning calorimetry (DSC) and to afford bulk thermosets after curing. Physico-chemical and thermo-dynamical characterizations were finally performed to compare these materials.

EXPERIMENTAL

Materials

Acrylic acid, triphenylphosphine, hydroquinone, octylamine, 2-hydroxyethyl acrylate, butyl acrylate and benzophenone were purchased from Sigma-Aldrich. Ethyl acetate (EtOAc) was purchased from VWR. Epoxidized linseed oil (Vikoflex 7190, EEW = 167 g/eq.) was obtained from Arkema. Priamine 1071 was obtained from Croda. PPG-diamines (D2000, Mn ≈ 2,000 g/mol; D400, Mn ≈ 400 g/mol) were obtained from Huntsman. Meta-xylylenediamine (MXDA) was purchased from CeTePox. All materials were used as received. Deuterated solvent CDCl3 was obtained from Eurisotop for NMR studies.

Synthesis of Acrylated linseed oil

In a round bottom flask, epoxidized linseed oil (20g, 0.12 mol of epoxy functions), acrylic acid (34.52 g, 0.48 mol), hydroquinone (0.02 g) and triphenylphosphine

(0.2 g) are heated at 80°C for 24 h. The flask is covered with aluminum paper to avoid UV exposure. When the reaction was complete, the mixture was diluted with 200 mL of ethyl acetate and washed three times with 100 mL of saturated sodium bicarbonate (NaHCO3) solution and one time with brine solution (100 mL). The organic layer was dried with anhydrous magnesium sulfate and filtered. The solvent was extracted under vacuum and the pure product was obtained as yellow viscous oil (quantitative yield).

RESULTS AND DISCUSSION

AELO synthesis

Acrylated epoxidized linseed oil was synthetized by ring opening reaction of epoxides with acrylic acid with quantitative yield. The conversion of epoxides into acrylate functions was monitored by 1H-NMR analysis. The disappearance of epoxy protons at 2.84-3.21ppm was followed by appearance of acrylate functions at 5.85, 6.10 and 6.40 ppm. The number of acrylate functions was characterized by NMR titration with an internal standard. The measured value of AEWexp for AELO is 380 g/eq. The theoretical value of AEWth is 239 g/eq. (EEW of ELO + Macrylate = 167 + 72 = 239). The activation of epoxy functions with triphenylphosphine causes also side-reactions such as epoxy homopolymerisation or epoxy-hydroxyl reactions. Less epoxy functions are thus available to react with acrylic acid addition. These side-reactions are responsible of oligomerisation and explain the difference of AEW between experimental and theoretical values. This can also explain the presence of many peaks between 3 and 3.5 ppm in 1H-NMR spectrum. Epoxidized linseed oil has an average value of 6 epoxy functions per triglycerides molecule. Thus, around 3 epoxy functions per molecule are involved in these side-reactions. Oligomerisation has already been observed in such monomers by GPC measurements. Improvement in terms of reaction parameters might limit these side-reactions. Indeed, other type of catalyst such as chromium-based catalyst (AMC-2), are known to promote epoxy ring opening while minimizing epoxy homopolymerization. However, these catalysts are toxic. Moreover, in the present study, oligomerization is not a drawback since we target the synthesis of thermosets, therefore, we chose cleaner catalyst, despite side-reactions. AELO has also been characterized by FTIR spectra. The peaks at 1620, 1400 and 800 cm-1 correspond to the CH=CH2 vibration of acrylate function.

Four different amines have been chosen to synthetize thermosets by aza-Michael reaction). These amines were chosen for their difference in terms of reactivity and structure in order to obtain materials with various mechanical properties. Priamine 1071, D2000, D400 and MXDA present various substituents which could have an influence on the reactivity of the amine. Fatty acid dimer-based Priamine 1071 is a diamine with a long aliphatic chain, whereas D2000 and D400 have a poly(propyleneoxide) PPO structure.

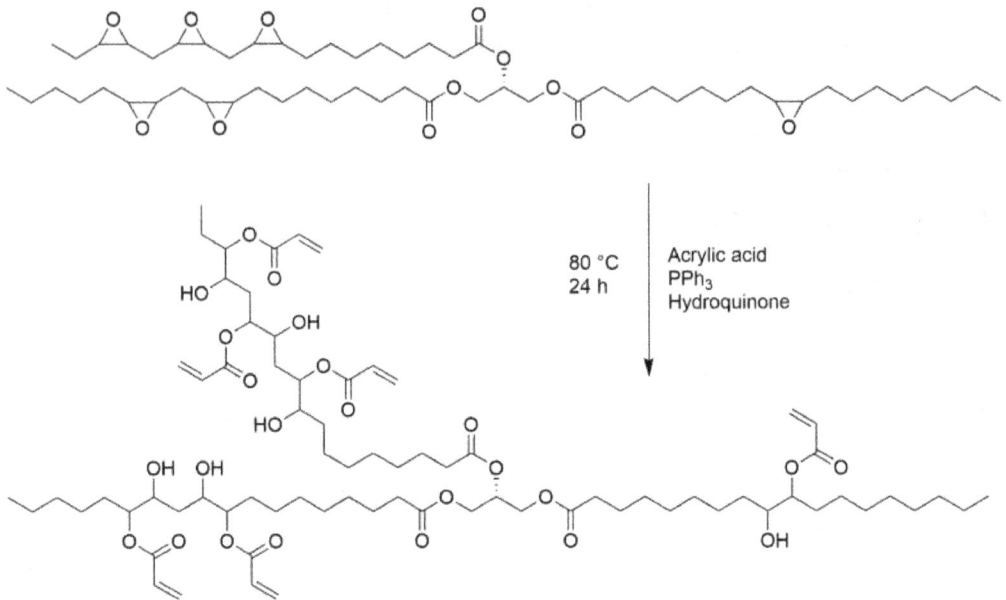

Figure 1. Acrylation of epoxidized linseed oil

Reactivity with different amines

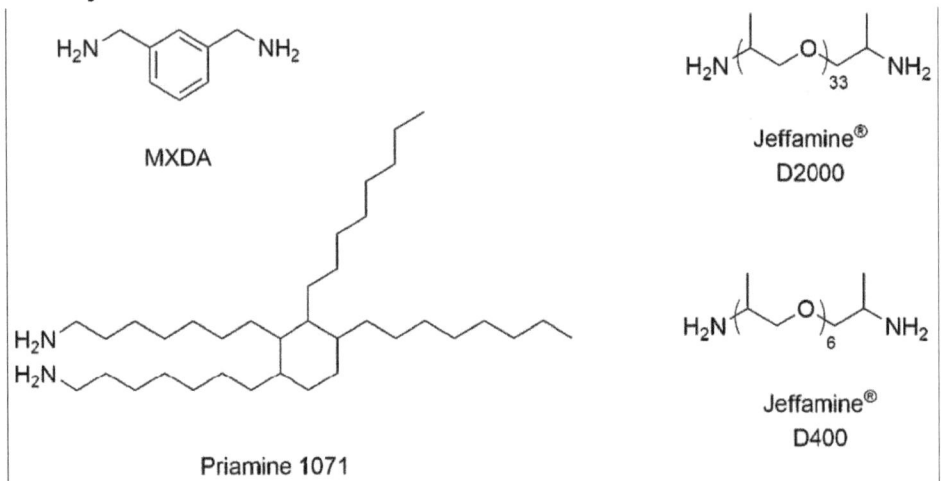

MXDA

Jeffamine® D2000

Jeffamine® D400

Priamine 1071

Figure 2. Amines structures

Finally the MXDA is an aromatic benzylic diamine. Kinetics of reaction between AELO and these amines were firstly studied by DSC analyses. Kinetics study with D400 was not performed due to its similarity with D2000 in terms of structure and reactivity. Long aliphatic diamines of different length were chosen in order to obtain soft materials whereas aromatic MXDA was used to increase the stiffness of final thermosets.

CONCLUSION

Aza-Michael reaction was performed between acrylated linseed oil and various amines to afford novel biobased thermosets. The role of hydroxyl in vicinal position to acrylate functions was firstly studied by FTIR and DSC analyses on model molecules which allowed highlighting the strong activation of acrylates by hydroxyl groups. The use of such structure is an interesting way to overcome conversion and kinetic limits. The attack of the secondary amine on acrylate is thus accessible without the use of catalyst, solvent or high temperature conditions. Different amines were used to synthetize AELO-based thermosets. Kinetics of curing was compared by DSC and FTIR measurements. With fast reaction time, Priamine 1071 affords fully biobased materials after simple curing at room temperature. In order to decrease the gelation time and facilitate processing, D2000, D400 and MXDA were also studied as curing agents. The steric hindrance of these amines limits the attack of the secondary amine. The primary amine is also slowed down for D2000 and D400 and thermal activation is necessary. The thermal and thermo-dynamical properties of cured materials were finally discussed. The structure of amines strongly affects the rigidity of the network. Materials with a large range of Tg are obtained from soft (AELO+D2000) to stiff (AELO+MXDA).

10
Chapter

VEGETABLE OIL-DERIVED EPOXY MONOMERS

INTRODUCTION

Since petroleum resources are ultimately limited, polymers based on vegetable oils are of great interest because they are renewable and could significantly contribute to a more sustainable development. Vegetable oils such as linseed and tung oil are drying oils, which can self-crosslink under atmospheric oxygen, have long been used in the coating industry. Semi-drying oils like soybean oil are of plentiful supply and therefore of relatively low cost, have also attracted great interest for the preparation of polymers or resins. In recent years, with the rising cost of fossil raw materials and environmental issues, polymers derived from soybean oil have demonstrated strong cost/performance competitiveness in many market applications. However, the ability to obtain structures of sufficient mechanical or thermal properties has remained a challenge.

For instance, direct radical or cationic polymerization of vegetable oils is structurally difficult due to the non-conjugated, internal double bonds and only viscous liquid polymers with low molecular weight are formed. On the other hand, polymers ranging from soft rubbers to hard plastics have been prepared by the cationic copolymerization of soybean oil blended with divinylbenzene (DVB). Styrene was added to reduce the heterogeneity of the crosslinked structures caused by incompatibility between monomers and the modulus of polymer was dependent on the styrene and, particularly, the DVB content.

Epoxidation of vegetable oils using peracids, such as Epoxidized soybean oil (ESO) and epoxidized linseed oil (ELO), is one of the most important and useful exploitations of double bonds since epoxides are reactive intermediates that are also readily converted to other functional groups through ring-opening reactions. Sheet molding compound (SMC) resins have been made from epoxidized soybean oil modified with unsaturated functional groups like acrylic acid or maleic anhydride where styrene was employed as a comonomer to reduce the viscosity of the resin. The SMC was obtained via common radical polymerization fashion. Allyl alcohol ring-opened ESO has been copolymerized with maleic anhydride (MA) to prepare thermosets by esterification and free radical polymerization. The resulting glass transition temperatures (Tg) and mechanical strengths were dependent on the loading of MA. Soy based polyols derived from ESO have also been widely used to produce polyure - thanes that are comparable in many aspects with polyurethanes obtained from petrochemical polyols.

ESO can be crosslinked into thermosetting polymers by various curing agents. However, due to lower oxirane content and sluggish reactivity of the internal oxirane, the cured ESO polymers normally have low crosslinking density. Poorer thermal and mechanical properties result from both partially unreacted ESO and saturated fatty acid (FA) chains that reduce reactivity and self-plasticize. Most ESO industrial uses are thus limited to nonstructural, additive applications such as plasticizers or stabilizers for poly (vinyl chloride), oil-base coatings with low strength requirements. Though the mechanical strength of cured ESO can be improved with the addition of nano-reinforcements, or fiber reinforcement, an inherently low Tg inevitably limits practical applications because Tg for a polymer must be appropriately higher than the temperature of its intended work environment to serve as a useful plastic. When used as a matrix material in composites, the resin state is desired to be rigid/glassy, i.e., below its Tg, to effectively transfer energy to fibers.

ESO has a moderate viscosity so ESO or their derivatives can be used as reactive diluents for the partial replacement of diglycidyl ether of bisphenol A (DGEBA) resins, which are relatively high viscosity liquids or solids, to decrease the overall cost and improve the processability. Generally, the mechanical strengths and thermal properties of ESO blended resins are not comparable to those of pure DGEBA epoxy resins, while their toughness can be better due to the introduction of a two phase structure. However, due to the inhomogeneous structure, ESO is not as efficient in reducing the viscosity of epoxy resin compared to most petroleum based reactive diluents. A further increase in the ESO concentration inevitably leads to a significant decrease in performance of cured resin. There are few reports of high ESO replacement because low oxirane content and the unreactive saturated component of ESO both lead to a low crosslink density upon cure and a poor miscibility exists between ESO and the DGEBA. There is an especially large difference in the reactivity of the internal oxirane in ESO and terminal oxirane in DGEBA and, as we will show, heterogeneous structures form during the curing reaction that leads to a phase separated materials of poorer mechanical /thermal performance.

More reactive terminal epoxy derived from chlorinated ESO has been reported and used as a matrix with DGEBA for glass fiber composites. The dehydro chlorination under alkaline conditions will hydrolyze ester groups of triglycerides, even at room temperature. A triglyceride with terminal epoxy has been synthesized from 10-undecenoic acid and successfully used in epoxy-amine curing, whereas 10-undecylenic acid, a derivative of castor oil, has only one terminal double bond so the epoxidized triglyceride ester of 10-undecylenic acid has a lower oxirane content compared to ESO. Largescale production also seems impractical. Only those oils of poly-unsaturated FA content, especially soybean or linseed oils that can produce dense oxirane functional resins are capable to produce satisfactory properties. Epoxidized vegetable oils (EVO) of low oxirane values either are not reactive or impart waxy, non-curing properties to the resin system.

Vegetable oils contain several actives sites amenable to chemical modification. The double bonds in FA chains and the ester groups in the glyceryl part are the most important. These active sites can be used to introduce reactive groups. ESO and aforementioned derivatives are most focused on the modification of FA chain. On the ester side, epoxidized methyl oleate, epoxidized methyl soyate, epoxidized allyl soyate, epoxidized sunflower oil biodiesel and linseed oil epoxidized methyl esters have been shown to have lower viscosity and more reactive compared to their ESO or ELO counterparts.

A caveat in the addition of functional groups, such as unreactive methyl or reactive allyl through transesterification, is a potential decrease of crosslinking density and final properties of cured resins upon breaking the oligomeric triglyceride structure. FA chain ends at the ester become pendant after transesterification and are dependent on crosslinking to build molecular weight. Esters of saturated FAs may only behave as plasticizers. Novel epoxy compounds such as epoxidized sucrose esters of fatty acids have been synthesized and crosslinked to prepare polyester thermosets. High modulus polymer was achieved due to the well-defined compact macromolecular structures and high oxirane functionality. Some applications may be hampered by their high viscosities.

Modified ELO synthesized through Diels–Alder reaction of dicyclopentadiene or 1, 3-butadiene with linseed oil have been reported. The modified ELO resins still possessed internal oxirane and thus are more suitable for cationic cure. End users still seek economical bio-based epoxies that are competitive with petroleum-based epoxies. Vegetable oils generally have variable levels of saturated FA content, for example, soybean oil normally has about 15% saturated FAs (~ 4.0 % stearic and ~ 11% palmitic) that varies with plant variety, growing regions, and weather. Saturated FAs have no functional groups within the FA chain that then act as dangling chains, low in reactivity, to plasticize the final polymer. The saturated chains are detrimental to the final properties of polymers.

To improve reactivity and to increase hydroxyl number of soy based polyols, regionally selective enzymatic hydrolysis has been attempted to liberate saturated

FAs, which were then removed by alkaline washing. Total removal of saturated components is difficult and is also accompanied by partial hydrolysis of unsaturated FA esters. Conversion of oil triglyceride into free fatty acid (FFA) or FA derivatives allows separation of unsaturated and saturated components on the basis of solubility through crystallization. The degree of unsaturation of FFA considerably changes the melting point and thus separation of mixtures of saturated and unsaturated FFAs can be readily achieved by proper choice of organic solvents and temperatures.

In this research, EGS were synthesized and examined. The goals were to remove and assess the role of the plasticizing effect of saturated components, to increase and assess the role of the oxirane content, and to minimize viscosity toward developing either a capable reactive diluent for commercial epoxy or a new commercial epoxy resin of its own right. The study gave us the opportunity to study how saturated component, oxirane type and oxirane content translate into curing, thermal and mechanical properties. We hypothesized that EGS as the ester of a terminal oxirane group (glycidyl), which is then readily accessible to nucleophilic attack, should further enable reactivity compared with the currently standard, commercial ESO and consequently reduce the molecular size and facilitate removal of the saturated FA components. We thus proposed to increase oxirane content. The goals and resin design were intended to provide a dense, intermolecular crosslinking structure and yield a more consistent thermosetting resin material with improved properties.

EXPERIMENTAL

Materials

Refined, food grade soybean oil (Great Value™, Wal- Mart, and Bentonville, AR, USA) was purchased. Linseed oil was purchased from Archer Daniels Midland Company (Red Wing, MN, USA). The major FA distributions reported for soybean oil and linseed oil. ESO was obtained from Union Carbide Corporation (Danbury, CT, USA). ELO was obtained from Arkema, Inc. (Philadelphia, PA, USA). Acetone, allyl alcohol, epichlorohydrin (EPCH), methylene chloride, methanol, meta-chloroperoxybenzoic acid (MCPBA), potassium hydroxide, sodium carbonate, sodium bicarbonate, sodium hydroxide, sodium sulfite, and anhy-

Table 1. Fatty acids profile in vegetable oils

Fattey Acid (x:y)	Palmitic (14:0)	Stearic (18:0)	Oleic (18:1)	Linoleic (18:2)	Linolenic (18:3)
Soybean oil [%]	11	4	23	53	8
Linseed oil	5	4	19	15	57

drous sodium sulfate were purchased from Fisher Scientific (St. Louis, MO, USA). Cetyltrimethylammonium bromide (CTAB), boron trifluoride mono - ethyl amine complex (BF3-MEA), 2-ethyl-4-methyl - imidazole (EMI), hydrochloric acid and 4-methyl- 1,2-cyclohexanedicarboxylic anhydride (MHHPA) were purchased from Aldrich (St. Louis, MO, USA).

Commercial DGEBA was supplied by Momentive (Deer Park, TX, USA) with trade name EPON™ Resin 828. Mold release agent Chemlease® 41-90 EZ was purchased from Chem-Trend, Inc. (Howell, MI, USA)

CHEMICAL CHARACTERIZATION

Infrared spectra (IR) were measured with a Nicolet Nexus 470 E.S.P. spectrophotometer (Waltham, MA, USA). 1H NMR spectra were obtained on a Varian INOVA 400 MHz spectrometer (Palo Alto, CA, USA) using d6-DMSO as solvent. Iodine value was assessed using ASTM Method D5554-95. Oxirane oxygen value was measured using AOCS Method Cd 9-57.

SOAP AND FREE FATTY ACID PREPARATION

Free fatty acids were made via acid neutralization of soap. Vegetable oil and water mixture (800 g, 50:50) was reacted with sodium hydroxide solution (200 g, 30 wt%) at 60°C for 4 hr to generate soap and then acidified with sulfuric acid (270 g, 30 wt%) to pH<2. The lower aqueous layer including sodium sulfate and glycerin was separated, washing the top FFA layer using 60°C water. Finally the liquid organic FFA layer (339 g) was dried using anhydrous sodium sulfate. The iodine value of the soybean FFA was 133.

Freshly prepared FFA was dissolved in acetone based on the weight ratio of 1:6 and then purged with nitrogen gas, cooled to –20°C for overnight. The formed crystals were removed by vacuum filtration. The procedure could be repeated several times until no further crystals were generated. For these studies, four times filtration were performed resulting in an iodine value for refined unsaturated soybean FFA of 150.

To a FFA/acetone solution (500 g) of weight ratio of 1:10, 110% of stoichiometric sodium hydroxide solution (18 mL, 10 M) based on amount of FFA (average molecular weight treated as 278 g/mol) was added dropwise. The neutralization reaction was continued for 4 hr under nitrogen gas to prevent air oxidation of the soap. The soap powder was readily filtered by vacuum filtration and then dried at 110°C for 1.5 hr.

GLYCIDYL ESTERS OF EPOXIDIZED FATTY ACIDS PREPARATION

Dry soap (302 g) and EPCH (925 g) were heated to reflux. Phase transfer catalyst CTAB (7.3 g) at 2 equivalent-% per equivalent soap was then added. Reflux was continued for 30 min, cooled and centrifuged, the clear solution was decanted to a flask. Excess EPCH was removed using in vacuo rotary evaporation. Oxirane oxygen value of prepared glycidyl ester was 4.4% (theoretical value of 4.7% for glycidyl oleate). Glycidyl ester (341 g) and sodium carbonate (64 g) were mixed with methylene chloride (200 ml). MCPBA (367 g, 75 wt%) dissolved in methylene chloride at 0.1 g/ml concentration was added dropwise at a reaction temperature below 15°C and then reacted for 4 hr to complete epoxidation. The reaction mixture was washed with 10 wt% sodium sulfite (200 g) and then by 10 wt% aqueous sodium bicarbonate (150 g). Methylene chloride was removed by in vacuo rotary evaporation and the product EGS (345 g) was dried over anhydrous sodium sulfate. Linseed oil based glycidyl esters of epoxidized fatty acids (EGL) were also prepared based on the above mentioned procedure. For EGSS/ EGL-S, saturated FFAs were not removed and remain in EGS/EGL. For EGS-P/EGL-P, FAs were partially epoxidized. Soybean oil based epoxidized methyl ester (EMS) and epoxidized allyl ester (EAS) were formed by standard alkaline transesterification with the corresponding alcohols and then epoxidized by MCPBA, e.g., potassium hydroxide (2.2 g) was first crushed and dissolved in allyl alcohol (260 g), then poured into soybean oil (220 g).

Mixtures were heated under reflux condition for 4 hrs. Workup included potassium hydroxide discharged by the addition of concentrated hydrochloric acid (3.9 g, 37 wt%), removal of the excess allyl alcohol using in vacuo rotary evaporation, washing of the allyl esters of soybean oil four times with distilled water to remove glycerin, salt, and any residual allyl alcohol, and then drying with anhydrous sodium sulfate and filtration to remove the sodium sulfate. The method for epoxidation of allyl esters of soybean oil by MCPBA is the above-mentioned method for epoxidation of glycidyl esters.

PREPARATION OF GLYCIDYL ESTERS OF EPOXIDIZED FATTY ACIDS

The synthetic route to EGS, generalized for oleic acid showing the process for a soybean triglyceride. Preparation of mixed FFAs from triglyceride is straightforward and well-developed. Methods of low temperature crystallization to remove the unsaturated FFAs are also well documented. Most unsaturated FFAs are soluble in most organic solvents at temperature above 0°C while the saturated FFAs, which have higher melting points than unsaturated FFAs, are prone to form crystals/precipitates at low temperature in solvents like acetone or methanol. Although trace amounts of saturated FFAs remain unavoidably in the unsaturated FFAs after low temperature crystallization, further removal of saturated FA components was achieved after synthesis of glycidyl ester or EGS because glycidyl esters, or the epoxidized glycidyl esters, of unsaturated FAs are each liquid at room temperature and much lower in melting point than glycidyl esters of saturated FAs. The unsaturated esters are poorer solvents for saturated carbon chains, which are then more easily precipitated at room temperature. Although no FFA component analyses, like chromatography, were performed in this research, we believe the saturated components were minimized after three precipitations.

Acetone was used as a low boiling, recoverable solvent to prepare soap. A slight excess of NaOH and higher concentration was preferable when preparing soap from FFA because unsaturated FFAs prone to dissolve in acetone rather than react with base. Unsaturated FFA soaps are more soluble in water.

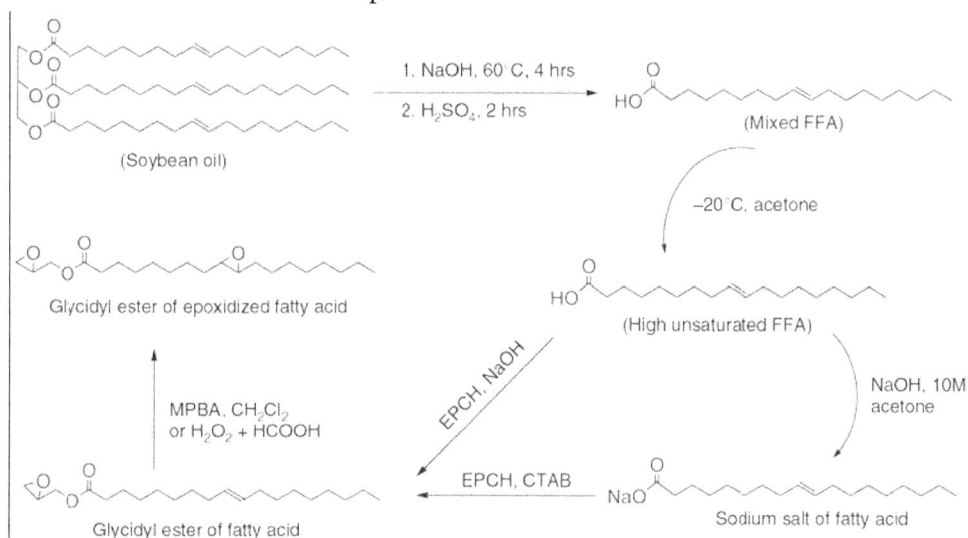

Figure 1. Synthetic route to EGS. (Vegetable oil and FAs are shown as simplified structures containing only oleic acid though they also contain other FAs. See Table 1)

Carefully dried and finely powdered soaps resulted in greater yields of glycidyl esters of FAs.

A low solubility of soap in EPCH suggested that a phase transfer catalyst would be useful to accelerate the reaction. With CTAB catalyst, the consumption of soap was completed within half an hour under reflux condition. Glycidyl esters can also be prepared directly from FFA in EPCH medium but the yield and purity were lower than obtained by the soap process. The epoxidation of glycidyl ester was carried out using MCPBA or in situ generated performic acid. The former was more efficient. Due to the low solubility of MCPBA in methylene chlo- ride, large amounts of recoverable solvent was required for the epoxidation.

In this study, the solubility parameter of each resin was calculated based on Hoy or van Krevelen model. All compounds structures are listed.

In ESO or EGS, the epoxy crosslink sites in the FA chains are located at the 9th and 10th carbons in the oleic acid and could be also at the 12nd and 13th carbons in linoleic acid, which leave the rest of the chain up through 18th carbon as an ineffective chain end in the crosslinked polymer.

EGS

DGEBA
(EEW = 186)

MHHPA

ESO
where R is epoxidized fatty acid chain

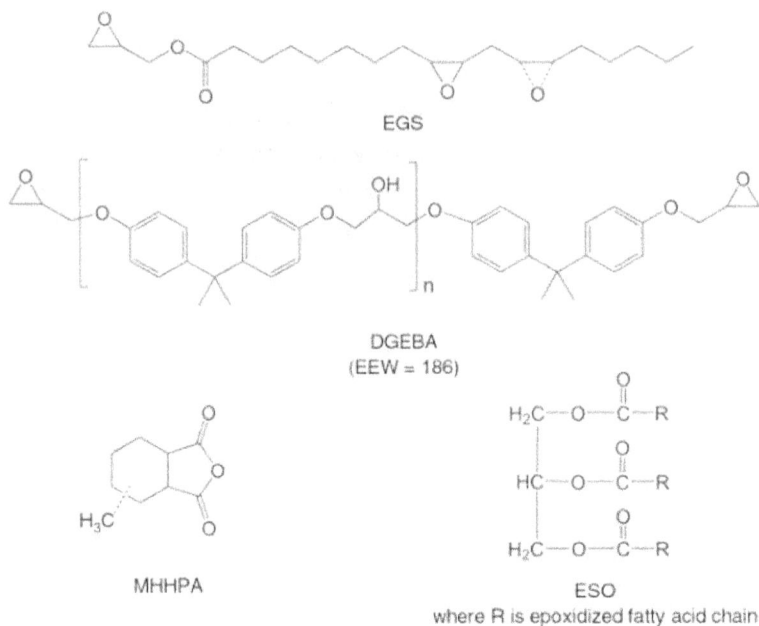

Figure 2. Compounds structure used for solubility parameters calculation

Furthermore, the presence of saturated palmitic or stearic acids in ESO triglyceride structure also behave like pendant chain , so the inactive parts, e.g., pendant chains and saturated FAs, in ESO and EGS constitute 34.8 and 18% of total mass, respectively.

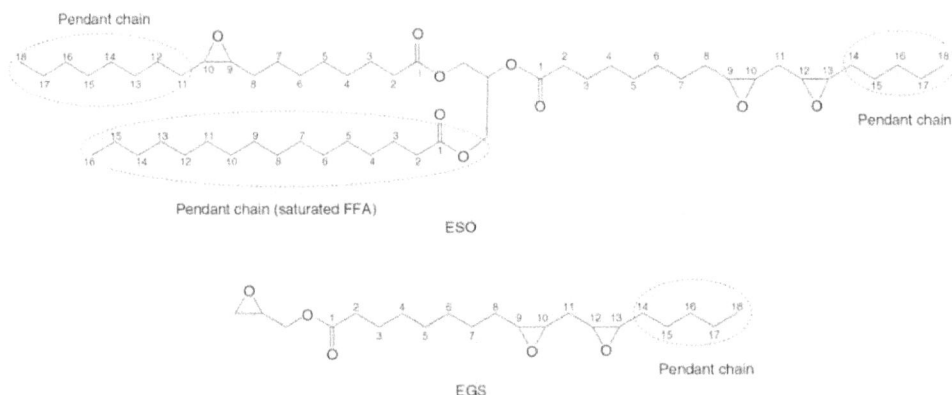

ESO

EGS

Figure 3. Schematic representation of pendant chain in ESO structure (epoxy moieties in ESO/EGS and methane moiety in glycerol part of ESO are the crosslink sites)

CONCLUSION

Bio-based epoxy resins, glycidyl ester of epoxidized fatty acids, were produced from soybean or linseed oils with a reduced saturated FFA fraction content. The products were characterized and showed high oxirane contents that were more reactive than ESO or ELO, which was shown to directly impact polymer homogeneity and glass transition temperature.

Epoxy monomers from other vegetable oil sources such as canola, palm, corn, etc., could be fabricated in similar fashion and have similar properties and curing behaviors providing that saturated fatty ester chains are similarly removed. The vegetable oil based epoxy resins displayed glass transitions that appear to be mostly a function of oxirane content but with additional influences of glycidyl versus internal oxirane reactivity, pendant chain content, and chemical structure and presence of saturated components. Generally, higher oxirane contents (epoxy functionality) lead to higher glass transition temperatures whereas reduced epoxy functionality, non-glycidyl FFA esters, and greater pendant chain contents lead to lower glass transition temperatures. In blends with DGEBA, monomers with only less reactive internal epoxies led to a more heterogeneous polymer structure compared to monomers possessing the more reactive glycidyl group and improved polymer homogeneity, in cure and structure. The inherent, long chain aliphatic structure of these thermoset monomers limits polymer glass transition temperatures compared to commercial, aromatic based epoxy monomers (DGEBA) but our data provide a clear trend and role of oxirane content.

The EGS blends with DGEBA were cured by MHHPA and their thermosetting polymer Tg's measured in comparison to control ESO blends with DGEBA, which were polymerized in similar fashion. The EGS polymers displayed improved Tg's and mechanical properties compared to their ESO counterparts and, in addition to an inherently low viscosity and efficient viscosity reduction, should therefore be more attractive as a reactive diluent.

For instance, EGS derived from renewable sources could further enable defect-free fabrication of complex, shaped epoxy composites for structural com- posite applications. Our data show ESO produced less homogeneous polymers when blended with DGEBA epoxy that resulted in thermal cure, thermal property, and mechanically inferior materials compared to the more compatible EGS epoxy resin and blends. The compatibility and superior properties arise from the removal of saturated pendant chains, addition of the glycidyl structure, and larger internal oxirane content inherent of EGS.

11
Chapter

LIGNIN-STYRENE-BUTYL ACRYLATE BASED COMPOSITE

INTRODUCTION

Lignin, in conjunction with cellulose, is one of the most abundant biopolymers in nature. Lignin is obtained as a by-product from the wood pulp during the paper fabrication. Due to its chemical structure, based in coniferyl, coumaryl, and sinapyl monomers randomly distributed and crosslinked, lignin has limited use in industrial scale processes. Lignin is a natural polymer based in coniferyl, coumaryl, and sinapyl monomers, randomly distributed and crosslinked. It is an amorphous material and hydrophobic branched which has recently been used for industrial applications, in a variety of alternatives.

During the last decades, a large number of studies have been presented on the potential use of lignin at the industrial level for the production of various materials, mainly as binder and dispersant, although lignin is a large source of aromatic compounds. Kasko et al. presented a review on the strategies to use lignin, where most of the research focuses on the methods developed for the synthesis of polymers from this biopolymer and its derivatives, giving a great value to the environmental benefit that entails the use of lignin as a raw material. Other studies have focused on the application of lignin as a nanofiller or in storage energy.

On the other hand, the study of the synthesis of compounds and nanocomposites through the grafting of lignin with acrylic monomers has been deepened, which have served as the basis for studying the compatibility and reinforcement effect of lignin blend whit different monomers, as well as the analysis of the compatibility of mixtures of lignin with epoxy resins. The research on the preparation of lignin with various polymeric materials with different applications has been studied from the viewpoint of matrix reinforcement with fibers or particles of different morphology. The nature of the matrix varies from a synthetic polymer to a natural one, depending on the application. It has been found that both the polymeric matrix and the reinforcement play an important role in the determination of the physicochemical properties of the composite materials in general. Fatehi, et al. reported the modification of a polymer of lignin-g-styrene and its application for wastewater treatment and water purification, by introducing a sulfonate group in the polymer matrix that modifies the anionic charge.

Under this context, we present the synthesis of a new compound based on lignin, styrene, and butyl acrylate, obtained by a bulk free radical polymerization. In the mechanism suggested in Scheme 1, styrene is added to the –OH groups of lignin through a free radical reaction. To understand the relationship between the lignin amount and the composite properties, the thermal properties, the contact angle, and the copolymer molecular weight, with and without lignin, was studied. The purpose was to deepen in the composite behavior to be able to visualize some potential applications; for example, as an adsorbent material to remove water pollutants, for synthetic wood, or for the absorption of hydrocarbons spills in water.

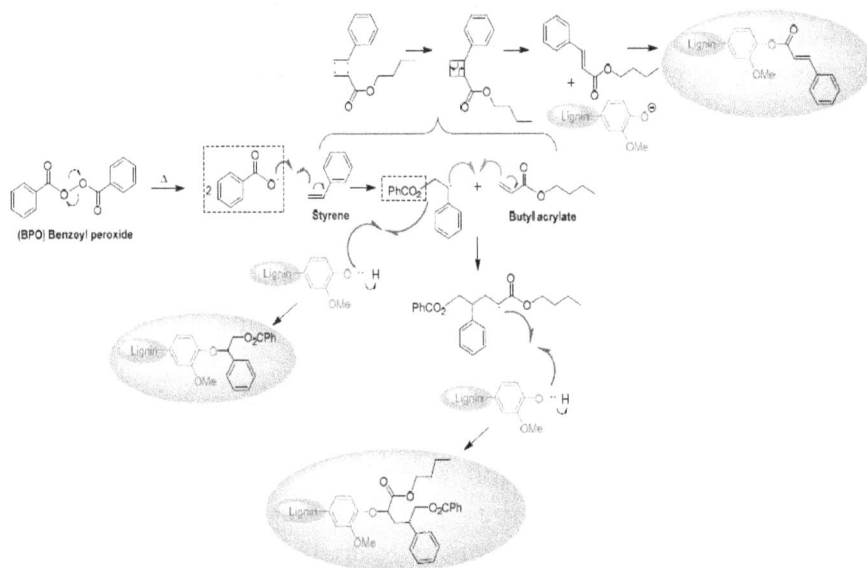

Scheme 1. Reaction mechanism for the chemical interaction between lignin and styrene-butyl acrylate copolymer.

EXPERIMENTAL

Materials

Lignin Kraft (98%, Sigma Aldrich, St. Louis, MO, USA) was dried in an oven for 24 h at 110 _C to remove moisture; the main characteristics of lignin are Mw 28,000 g mol-1 and Mn 5000 g mol-1. Styrene monomer (98% DEQ, México), butyl acrylate (97% Sigma Aldrich), and benzoyl peroxide (99% Sigma Aldrich). All reagents were used without further purification.

COMPOSITE SYNTHESIS

The composite synthesis was performed by bulk free radical polymerization, mixing the defined amount of styrene (S), butyl acrylate (BA) and lignin, keeping the reagents total amount of 4.0 g. Benzoyl peroxide (C14H10O4) was fed at 1 wt.% with respect to the total amount of the reaction mixture. A total of five experiments were performed. The relationship between lignin and styrene used. The polymerization time for all the experiments was 2 h, at 90OC, and maintaining a stirring speed of 1200 rpm. At the end of the established time, the materials were cooled rapidly in an ice bath (quenching). The products resulting from the experiments were identified as EBA for the material without lignin and LEBA for the materials containing lignin.

Table 1. Lignin and styrene quantity present in composites, with butyl acrylate (BA) constant in 14 wt.%.

EBA	Lignin [wt.%]	Styrene [wt.%]	BA/St ratio
LBA	0	86	0.16
LEBA5	5	81	0.17
LEBA10	10	76	0.18
LEBA15	15	71	0.20
LEBA20	20	66	0.21

RESULTS AND DISCUSSION

The butyl acrylate (BA) monomer was included to impart slight flexibility and to prevent obtaining a brittle composite. As noted, after 1.3 h the residual styrene was less than 0.04%; after this time, the conversion reached a steady state, so the composite synthesis can be stopped. This method of synthesis has advantages over other reports that involve the incorporation of lignin in polymeric matrices, which require the use of solvents or a pre-treatment to lignin or are necessary longer synthesis time.

After this time, the conversion reaches a steady state, so the synthesis of the composite can be stopped at 1.5 h of reaction. Short reaction time implies energy savings when the compound is developed at an industrial level. According to this, the interaction of –OH groups of lignin with the copolymer during free radical polymerization increased (Scheme 2), occupying a higher proportion of this groups, which caused the decrement in the hydrophilic character associated with lignin.

In this sense, some research in the synthesis of lignin-based materials, reported that the increase of lignin in the polymer matrix influences the increment of the hydrophilic character, because lignin acts as filler. On the other hand, under the synthesis conditions proposed in this research it was shown that the increase in the amount of lignin (LEBA15 and LEBA20) had no significant effect on the hydrophobic character of the styrene-butyl acrylate system.

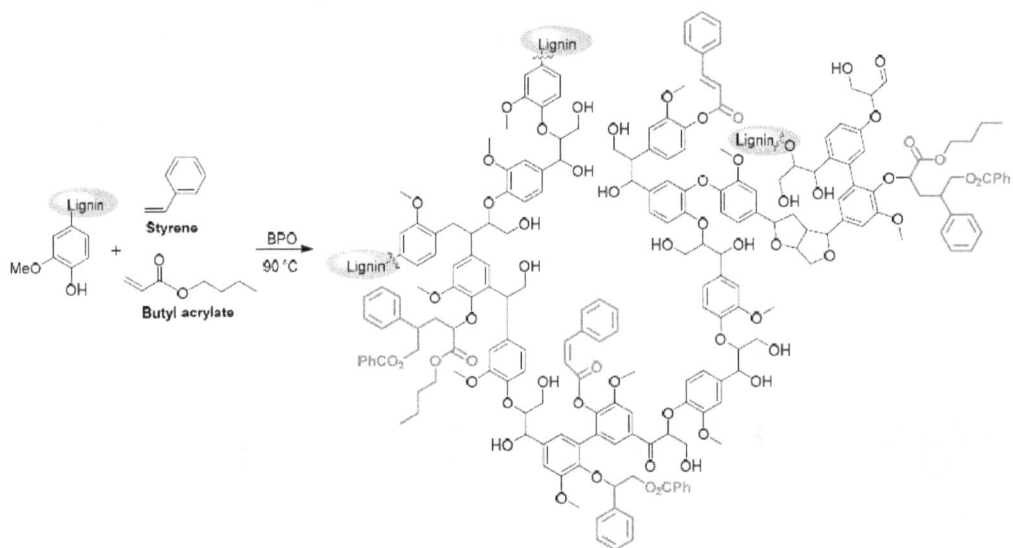

Scheme 2. Schematic diagram of the probable links in composites synthetized with lignin and styrene butyl acrylate

CONCLUSIONS

A novel composite was synthetized with lignin and styrene-butyl acrylate copolymer. Under the conditions of the composite formation, the results indicated that the –OH groups of lignin were the main site of bonding with the copolymer chains; this was evidenced by the reduction of the hydroxyl groups signals in infrared spectroscopy. Gel permeation chromatography showed that the atmosphere of nitrogen in the reaction medium did not contribute significantly to the inhibition of the free radicals during copolymerization. In addition, electron microscopy showed that the lignin was incorporated with the copolymer, losing its granular morphology and producing a smooth surface, which was related to FTIR observations. Respecting surface characterization, it was found that lignin loading modified the composites hardness and the hydrophilic-lipophilic character. The more hydrophobic character of LEBA15 and LEBA20 composites supported the assumption of the chemical interaction between the OH groups of lignin and the copolymer chains during free radical polymerization. According to the properties observed, it is possible to suggest applications for the composites; for example, based on the lipophilic nature, the composites could be used as adsorbent materials for the removal of hydrocarbon spills from water.

12

Chapter

BIOCOMPOSITES FROM LACTIC ACID THERMOSET RESINS

MATERIALS & METHODS

L-Lactic acid (88-92%, Sigma-Aldrich) was purified by using a rotary evaporator. Glycerol (99.5%, Fisher Scientific), allyl alcohol (≥98.5%, Sigma-Aldrich) and pentaerythritol (98%, Sigma-Aldrich) were used as received. Toluene was used as solvent (99.99%, Fisher Scientific) and methane sulfonic acid (98+%, Alfa Aesar) was used as the catalyst in the condensation reaction. Methacrylic anhydride (94%, Alfa Aesar) was used as the reagent for the end-group functionalization. Hydroquinone (99%, Fisher Scientific) was used as inhibitor during the end-group functionalization reaction. An unsaturated polyester resin (Polylite 444- M850, Reichhold AS) was used as a reference when evaluating the synthesized resin properties. This commercial resin is a medium reactive orthophtahlic type resin, which contains approximately 43 wt-% styrene. For the curing of the resins, two different initiators were used; methyl ethyl ketone peroxide (33%, BHP Produkter AB) for room temperature curing, and dibenzoyl peroxide (>98%, Kebo Lab) for curing at 150°C. N, N-dimethyl aniline was used as accelerator with the dibenzoyl peroxide.

Three structurally different resins were synthesized from lactic acid namely; glycerol-lactic acid resin (GLA resin), methacrylated allyl alcohol-lactic acid resin (MLA resin) and pentaerythritol methacrylated lactic acid resin (PMLA resin). All reactions were performed under nitrogen atmosphere.

1.2.1 GLA resin (glycerol-lactic acid resin) star-shaped oligomer of glycerol and lactic acid was prepared. Three oligomers with different chain lengths in the glycerol branches (n = 3, 7 and 10) were made in order to study the influence of the length of the chains. This was done by using 9, 21 or 30 moles of lactic acid for each mole of glycerol. The reactants were diluted in 50 g of toluene and 0.1 wt-% methane sulfonic acid was used as catalyst.

All components were placed in a three-necked round bottom flask equipped with a magnetic stirrer, nitrogen inlet and a Dean-Stark azeotropic distillation apparatus. The flask was heated for 2 h in an oil bath with a set temperature of 145°C under constant stirring. The water produced in the reaction was collected by azeotropic distillation. After the initial reaction, the temperature was raised to 165°C for further 2 h, and finally increased to 195°C for 1 h. In the second stage, the remaining reactant solution was cooled to 110°C and maintained at that temperature. Hydroquinone (0.1 wt-%, used as a stabilizer) was then added into the reaction mixture. The end-functionalization was done by adding dropwise during 4 h, 0.396 mole of methacrylic anhydride (n = 3) while being constantly stirred under nitrogen atmosphere. After the end-functionalization, the formed methacrylic acid and the remaining toluene present in the resin mixture were removed by rotavapor distillation at temperature of 60°C and 13 mbar pressure. The chemical reaction for the GLA resin synthesis is shown in Figure 1.

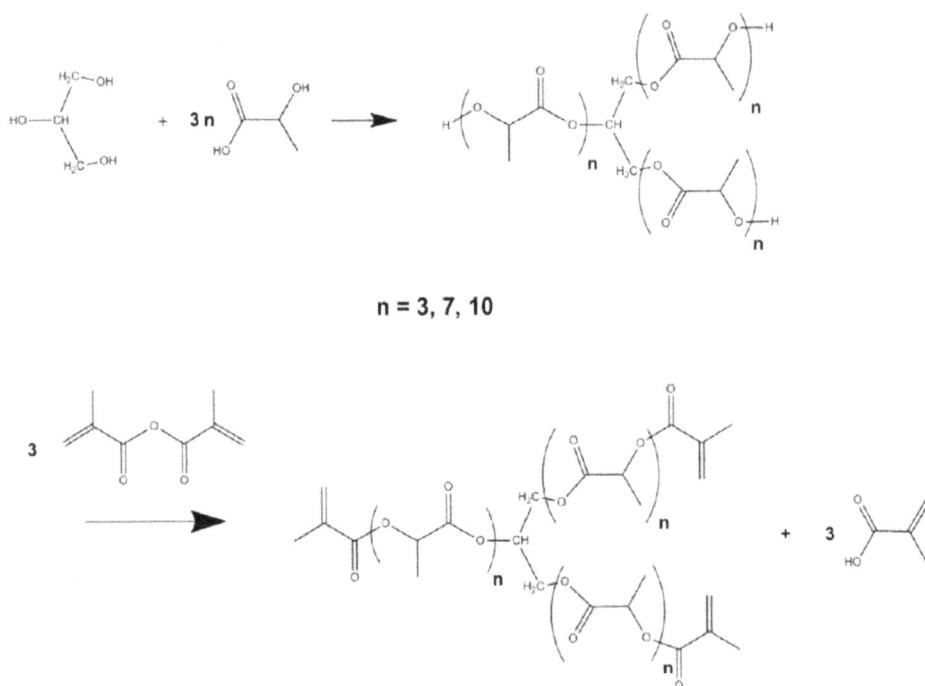

n = 3, 7, 10

Figure 1. Synthesis reaction for methacrylated glycerol-lactic acid resin (GLA resin) with n = 3, 7 and 10.

MLA RESIN (METHACRYLATED ALLYL ALCOHOL-LACTIC ACID RESIN)

The synthesis was performed in two stages. In the first stage, an intermediate allyl alcohol lactic acid oligomer (ALA resin) was prepared from 1 mole of allyl alcohol and 5 moles of lactic acid. The reaction was done in 250 g toluene, and with 0.1

wt-% of methane sulfonic acid as catalyst. All chemicals were placed in a three-neck round bottom flask equipped with a magnetic stirrer, nitrogen inlet and a Dean-Stark azeotropic distilling trap. The flask was heated for 2 h in an oil bath with a set temperature of 145°C under constant stirring. The water produced in the reaction was collected by azeotropic distillation in the trap. After the initial 2 h, the temperature was raised to 165°C for further 2 h, and finally increased to 195°C for 1 h. In the second stage, the obtained ALA resin was reacted with methacrylic anhydride to end-functionalize the oligomer. The intermediate reaction mixture was cooled to 90 °C and a stabilizer of 0.2 wt-% of hydroquinone was added under stirring. Finally 1.1 moles of methacrylic anhydride were added drop-wise during 4 h into the reaction mixture while being constantly stirred under nitrogen atmosphere.

After the second stage, the formed methacrylic acid and the remaining toluene present in the resin mixture were removed by rotavapor distillation at temperature of 60°C and 13 mbar pressure.

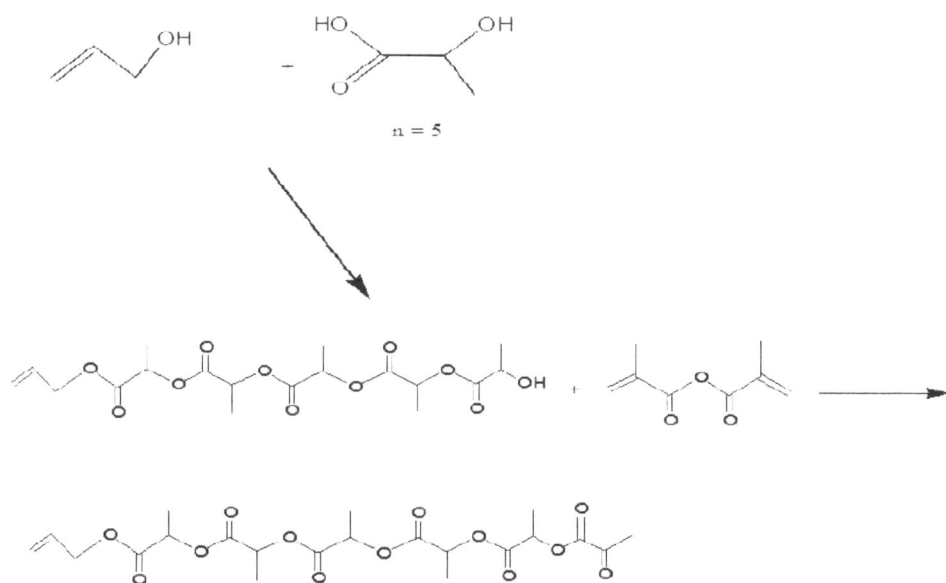

Figure 2. Synthesis reaction for allyl alcohol lactic acid oligomer resin (ALA resin) and methacrylated allyl alcohol-lactic acid resin (MLA resin).

PMLA RESIN (PENTAERYTHRITOL METHACRYLATED LACTIC ACID RESIN)

The synthesis was also done in two stages. The first stage was equivalent to the first stage used for MLA resin synthesis. In the second stage, four moles of the ALA resin were reacted with one mole of pentaerythritol. The reaction was done at 90°C and a stabilizer of 0.2 wt-% of hydroquinone was added into the stirred reaction mixture. Then an excess of 10 moles of methacrylic anhydride was added drop-wise during 4 h while being constantly stirred under nitrogen atmosphere. After the end-functionalization, the formed methacrylic acid and the remaining toluene present in the resin mixture were removed by rotavapor distillation at temperature of 60°C and 13 mbar pressure.

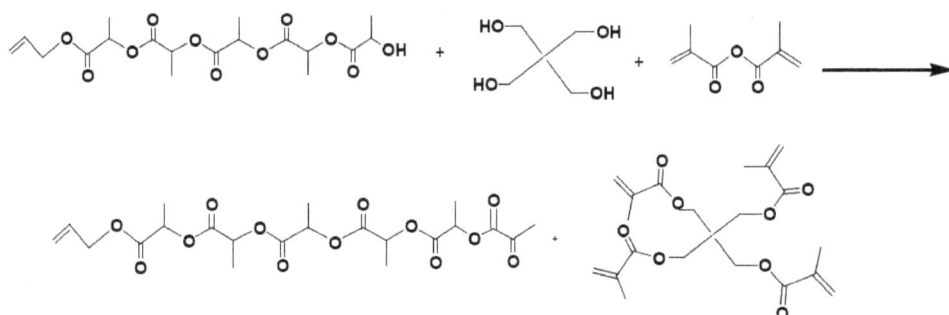

Figure 3. Synthesis reaction for pentaerythritol methacrylated lactic acid resin (PMLA resin).

The bio-based thermosetting resins demonstrated in this thesis have shown promising mechanical, thermal and rheological properties. They are of high renewable ratio, which encourages the use of renewable reactants and they can compete with many commercial polyester resins. It is a straightforward, relatively cheap methodology and can be up-scaled for industrial production.

Composites have been prepared from these bio-based resin reinforced with natural and regenerated cellulose fibre, which increases the renewability ratio of the composites produced. The mechanical properties of the composites are very good and they can be used in some applications to replace conventional polyester reinforced composites. Based on the aging properties of the composite produced, they can be used for indoor applications such as interiors of automobiles or as furniture applications. These bio-based thermoset resins can possibly be used with synthetic fibres for outdoor application which can reduce its nonrenewable ratio and give good mechanical properties. Considering the high renewable ratio of the bio-based resins, promising properties for making strong affordable, biodegradable and renewable resourced composite for many applications can be achieved.

13
Chapter

BIOMATERIALS FROM STARCH

75% of all organic material on earth is present in the form of polysaccharides. . An important polysaccharide is starch. Plants synthesise and store starch in their structure as an energy reserve. It is generally deposited in the form of small granules or cells with diameters between 1-100 µm. Starch is found in seeds (i.e. corn, maize, wheat, rice, sorghum, barley, or peas) and in tubers or roots (i.e. potato or cassava) of the plants. Most of the starch produced worldwide is derived from corn, but other types of starch such as cassava, sweet potato, potato, and wheat starch are also produced in large amounts. Most starch crops are very productive. Potato accumulates starch to approximately 75 % of the dry weight in the tubers with a yield up to 21 ton starch per hectare, while corn seeds consist of 65-80% starch by weight, with an average yield of 4.9 ton starch per hectare.

PLASTIC APPLICATIONS AND WASTE ISSUES

Plastic is the general term for a wide range of synthetic or semisynthetic polymerisation products. Plastics are used in a wide range of applications and the demand is still increasing every year. The first generation of commercial plastics was derived from cellulose nitrate and is known as celluloid.

Cellulose nitrate was first prepared by A. Parker in 1838, and celluloid was patented by J. Hyatt in 1870. While celluloid is derived from a natural polymer (cellulose), the oldest purely synthetic plastic is Bakelite, discovered by Baekeland in 1907. A dramatic increase in demand for plastics began after World War II, when polyethylene (PE) was invented. PE is a very versatile plastic because it can be shaped easily into various forms, for instance to be used in packaging and paper coatings.

Plastics are very attractive materials. They have a low density and can be shaped in thin films while maintaining good properties. The latter is important when using the material for packaging purposes to save weight, space, and energy during transportation of goods. Plastics have lower melting temperatures compared to glass and metals, and therefore need less energy to shape it into useful materials.

The production and consumption of plastics has increased significantly with a rate of almost 10% every year since 1950. In 2006, the worldwide plastics production has reached 245 million ton per annum. The largest application of plastics is for packaging purposes. About 29% of the total plastics produced in the USA, and 37% of the total plastics demand in EU is used as packaging materials. Important polymers used for packaging are polyethylene (HDPE and LDPE), polypropylene (PP), polystyrene (PS), polyvinyl chloride (PVC), polyethylene terephtalate (PET), and polycarbonates (PC). Plastics are also used for building materials and automotive, electrical, and consumer products.

Plastic waste, however, is causing serious environmental problems. It has a high volume to weight ratio and is resistant to degradation. Plastics have been polluting sea, soil, rivers, and lakes, threatening fishery, ship navigation, hydropower plants operation, irrigation, and other public works. Plastic litter is hazardous to wildlife, and accumulation of its residues in soil cause significant reductions in agricultural yields.

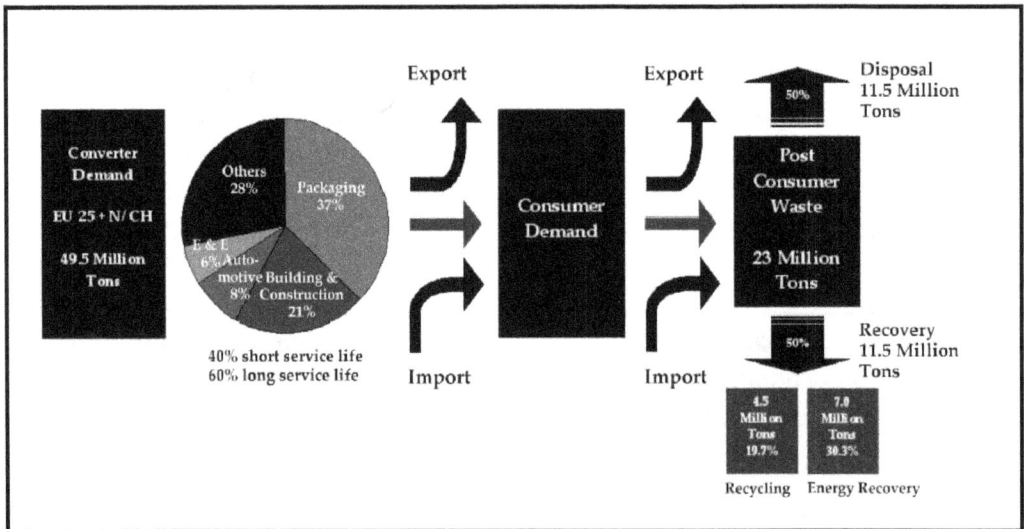

Figure 1 Plastic applications and waste treatment in Europe

The disposal of plastics materials in municipal solid waste (MSW) is a serious issue in many parts in the world. In the USA, the amount of plastics in MSW increased from less than 1% in the 1960 to 11.7% in 2006. The volume fraction of plastic in MSW is much larger due to the low density of plastics, and may be more than twice the weight fraction. In 1996, plastics waste was ranked as the second major source of MSW after paper and paperboard, consisting of 25%-v of the total waste. Recycling (part of) the plastics to reduce the amount of MSW also has limitations. It is not applicable for thermoset resins, and is only effective for single plastic sources or simple plastic formulations. Co-mingled plastics, which are usually found in waste streams, are not easy to recycle. In Europe (2006), only 20% of the plastic is recycled. Most of the waste is still disposed (50%) by landfill or incinerated to recover the energy (30%).

THE POTENTIAL OF BIODEGRADABLE PLASTICS

The application of biodegradable plastics could be an attractive solution for the problems related to the use of conventional plastics (vide infra). Biodegradable plastics are polymeric materials capable of decomposing when given an appropriate environment and sufficient amount of time. Biodegradable plastics have gained considerable interest since the 1980s. Nowadays new types of biodegradable plastics with improved properties and lower costs have been developed. Biodegradable plastic waste may be treated in composting facilities, together with food and yard waste as well as paper. It may also be treated in sewage sludge water treatment plants or buried in the soil. The considerable growth of interest in composting as a means to replace landfill due to the decreasing disposal spaces (especially in Europe) may also help the progress of biodegradable plastics development.

Biodegradable plastics may be classified into two general groups, biopolymers from nature (from plants, animals, and microorganisms) and biodegradable synthetic polymers. Common natural biopolymers are carbohydrates, proteins (present abundantly in plants and animals), and polyesters from micro-organisms. Biopolymers are inherently biodegradable, because they take part in nature's cycle of renewal.

BIODEGRADABLE PLASTICS FROM STARCH

Starch is a very attractive source for the development of biodegradable plastics. The price of starch in 2007 was about $ 0.45 per kg. This price is much lower than conventional plastics derived from oil, such as polyethylene (PE, € 1.2-1.4 or $ 1.8-2.4 per kg). Starch may become an attractive raw material for plastics in the future, because the price of oil based polymers may still increase due to the rise in the crude oil prices. Bastioli showed that nearly all biodegradable plastics available in the market are derived from starch, either from starch-based materials (slightly modified starch, alone or complexed with natural or synthetic biodegradable polymers) or from polylactic acid which originates from the fermentation of a starch feedstock.

Virgin starch is not suitable as a packaging material. It cannot be shaped in films with adequate mechanical properties (high percentage elongation, tensile and flexural strength) and is too sensitive to water. Consequently, starch must be modified, either by plasticization, blending with other materials, chemical modification or combinations of them, before they can be applied as biodegradable plastics.

STARCH MODIFICATIONS TO IMPROVE PRODUCT PROPERTIES

Several techniques may be applied to develop starch based biomaterials with improved properties. These are summarized briefly in the next sections.

Thermo plasticized starch

Virgin starch is brittle and difficult to be processed into articles due to its relatively high glass transition temperature (T_g, approximately 230°C), which is even above the thermal degradation temperature. . The brittleness is known to increase in time due to free volume relaxation and retrogradation. This problem is mainly caused by the presence of strong inter - and intra-molecular hydrogen bonds between the starch macromolecules.

Starch can be modified to obtain materials which melt below the decomposition temperature, and therefore are processable by conventional polymer processing techniques such as injection, extrusion, and blow moulding. The modified products are known as thermoplastic, destructed, plasticized, or melted starch. The modification involves break down of the starch granular structure by the use of plasticizers at high temperatures (90-180°C) and shear, which will result in a continuous phase in the form of a viscous melt. The thermo plasticization process will decrease the interactions of the molecular chain and destruct the structure of the starch. As the result, the semicrystalline structure of starch and its granular form are lost and the starch polymers are partially depolymerized, resulted in the formation of an amorphous mass.

There are several substances used as plasticizer for the preparation of thermoplastic starch (TPS), such as water and polyols (glycerol, glycol, sorbitol, sugars). The use of water as a plasticizer is not preferable, because the resulting product will be brittle when equilibrated with ambient humidity.

The use of other plasticizers (for example glycerol) results in a rubbery material, with better properties than virgin starch in various applications. Yu, et al discovered that the elongation of break of the thermoplastic starch is significantly improved by plasticization with glycol, glycerol, and hexylene glycol. The plasticized starch properties may be tuned by changing the temperature of processing, water content, and the properties and amount of plasticizers. For instance, the thermal properties of glycerol-plasticized starch are a function of water content. At intermediate water levels, phase separation may still occur. A biodegradation study according to ISO/CEN 14852 and ASTM D5209-92 standards on films made from starch–glycerol–water mixtures confirmed that the films are easily biodegradable.

Although thermo plasticization seems to be a promising method, TPS synthesized from polyol and sugar plasticizers have the tendency to recrystallize (retrogradation) after being stored for a period of time, which results in embrittlement. Another issue is the poor water resistance and low strength, which still limits their use. The plasticizers are also usually hydrophilic and can be washed out by water. Solutions to improve the properties of TPS are blending, or coating with hydrophobic polymers.

Cross-Linked starch

An anhydroglucose molecule of starch contains two secondary and one primary hydroxyl group. These hydroxyls can react easily with a wide range of compounds such as acid anhydrides, organic chloro-compounds, aldehydes, epoxy, and ethylenic compounds. Chemicals of these classes having two or more of the reactive groups may react with two or more hydroxyls of the starch molecules. The products are called cross-linked starches. Cross-linking results in a reduction of the solubility in water and to thickening, leading to higher viscosities. The cross-linked starches have found many applications, especially in the food, paper, textile, and adhesive industry.

Silva, et al studied the mechanical properties of films from maize starch cross-linked by sodium trimetaphosphate (SMTP). Higher processing temperatures generally led to higher cross-linking levels, resulting in an increase of the Young's modulus and tensile strength of the products and a decrease in elongation at break. The cross-linked products are therefore more rigid (and less elastic) materials than virgin starch.

Starch Esters

The development of starch esters started in the mid 19th century. Already in 1865, Schuetzenberger acetylated starch with acetic anhydride. Most studies dealt with the synthesis of starch esters of C1-C4 carboxylic acids, and particularly with acetic acid.

The simplest starch ester is starch formate (C1), synthesized by direct addition of formic acid to starch at room temperature. The esterification reaction is catalyzed by H+ and proceeds with the formation of water. Breakdown of the starch and the formation of low molecular weight products occur to a significant extent. As a result, the use of formyl esters of starch is much lower than that of acetate esters. The most popular starch ester is starch acetate. Three different types of starch acetates may be distinguished, differing in Degree of Substitution (DS). The DS is defined as the moles of substituents per mole of AHG units. Low DS products (0.01-0.2) are commercially available and used as food additives and in the textile industry. Medium DS starch acetates (0.3-1) are still soluble in water, while highly substituted starch acetate (DS of 2-3) is soluble in organic solvents. Substitution of the hydroxyl groups of starch with acetate groups makes the esters more hydrophobic than native starch. High DS starch acetate can be easily casted into films using organic solvents. The degree of acetylation of starch acetates can be easily controlled, allowing the polymer to be produced with a range of hydrophobicities. High DS starch acetates are thermoplastic materials suitable to be used as biodegradable plastics.

Several synthetic routes have been developed for starch esters. Esterification may be performed using acid anhydrides in aqueous media or organic solvents (pyridine, DMSO, xylene, DMF, or isopropanol) with acidic (hydrochloric or sulfuric acid) or alkaline (NaOH or triethylamine) catalysts. Starch triacetate has been successfully synthesized using acetic anhydride in combination with pyridine-gelatinized starch. Starch esters have also been synthesized using alkanoyl chlorides. Another attractive route involves the use of vinyl esters as reagents. The kinetics of the reaction between gelatinized aqueous potato starch and vinyl acetate was studied by De Graaf, et al,. Acetylation of starch in water and DMSO using vinyl acetate has been studied lately by Mormann and Al-Higari as well as Dicke.

Neutral and weak acid/ alkaline catalysts result in region selective substitution at the C2 hydroxyl groups of starch, while alkaline catalysts (such as carbonate, hydrogen carbonate, acetate, and phosphate) will result in C-2, but also C-6 and C-3 substitution.

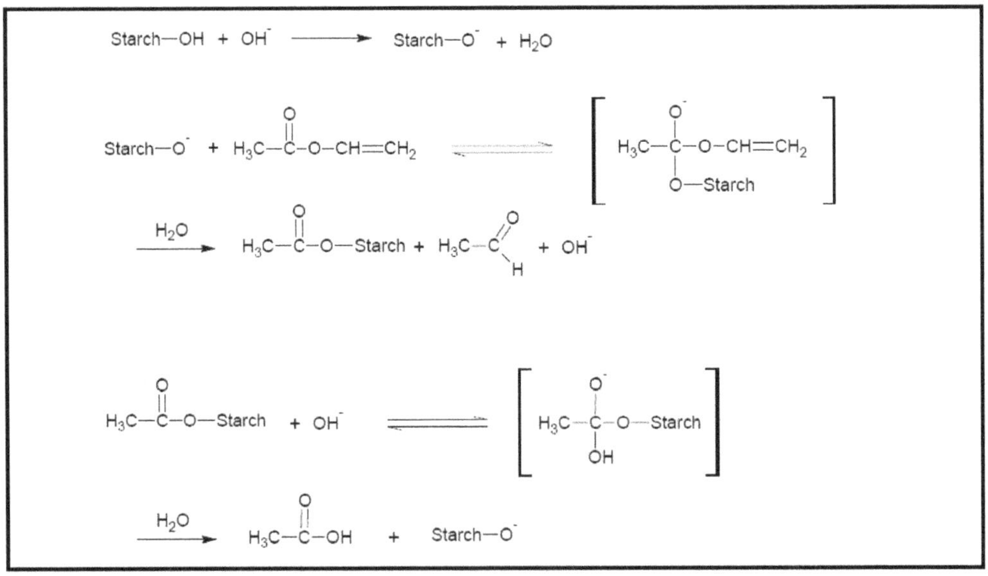

Scheme 1 Mechanism of starch acetylation using vinyl acetate

The synthesis of long chain fatty acid ester of starch has attracted much interest lately. The introduction of longer acid chain is expected to reduce the brittleness of virgin starch and to increase its hydrophobicity . Fatty-acid starch esters have been synthesized using fatty acid (octanoyl, dodecanoyl, octadecanoyl) chlorides. The fatty acid chloride reactants are, however, relatively expensive and rather corrosive. The use of methyl and glyceryl esters of fatty acid (in the absence of solvent) to synthesize starch fatty acid esters has also been studied, but only relatively low-DS (0.34-0.61) products could be obtained using this approach.

STARCH-BIOPOLYMER BLENDS AND GRAFT CO-POLYMERS

Blending of different polymers is an established method to obtain products with improved properties. However, polymers are rarely miscible with each other so that, in the simplest case of a binary blend, one component will be dispersed into the other. The degree of adhesion (binding) between the dispersed phase and the matrix is dependent on the molcular interactions between the two components and represents a crucial factor in determining the morphology of the blends and eventually the product performance. Two main methodologies are applied for the production of polymer blends. The first involves simple melt mixing of the two components for example by extrusion. By working at temperatures above the melting point or glass transition temperature of the two components, the latter are mixed together. If the right combination of chemical groups on the two components is present along the polymers backbone, a chemical reaction might take place upon processing (reactive extrusion). The second methodology for producing polymer blends involves the in situ polymerization of one component (thus originally present in the blend in monomeric form) in the presence of the second one. The classical example of such process is represented by the production of High Impact Poystyrene (HIPS) obtained by styrene polymerization in the presence of polybutadiene.

Although the in situ polymerization process is not as technologically straightforward and economically convenient as melt-mixing, it is frequently used in order to chemically graft the polymerized component on the other one (polystyrene on polybutadiene). As a result, the two components are chemically linked to each other, which in turn provide a very strong adhesion at the molecular level between the dispersed phase and the matrix. Despite this advantage, melt-mixing remains the preferred route to polymer blends mainly because of very practical reasons: low costs, availability of mixing equipment and no necessity to use any organic solvent (often employed for the in situ polymerization). However, in order to achieve also a good molecular adhesion between the phases by melt mixing, interfacial agents (e.g. compatibilizers) might be used. Their role is comparable to the one of a surfactant in emulsion formation, i.e. they locate themselves at the interface between the two components stabilizing the dispersion (most probably by a steric repulsion mechanism) and providing an improved adhesion at the interface. A suitable interfacial agent for the blends of two polymeric materials is a block copolymer for which the chemical structure of every block is the same (or very similar) to the one of the individual components to be blended.

Interfacial agents already available on the market can be used as such or can be produced upon mixing (in situ) by chemical reaction of the two components. In the past, there have been efforts to blend as well as to graft synthetic polymers onto starch. The synthesis of compatibilizer and its use for starch/ synthetic polymer blending has also been studied. The products have been synthesized in the lab as well as on industrial scale. The blending and grafting of starch with synthetic polymers is usually performed to achieve higher hydrophobicity and to improve the mechanical and thermal properties as well as to obtain cheaper and more biodegradable products.

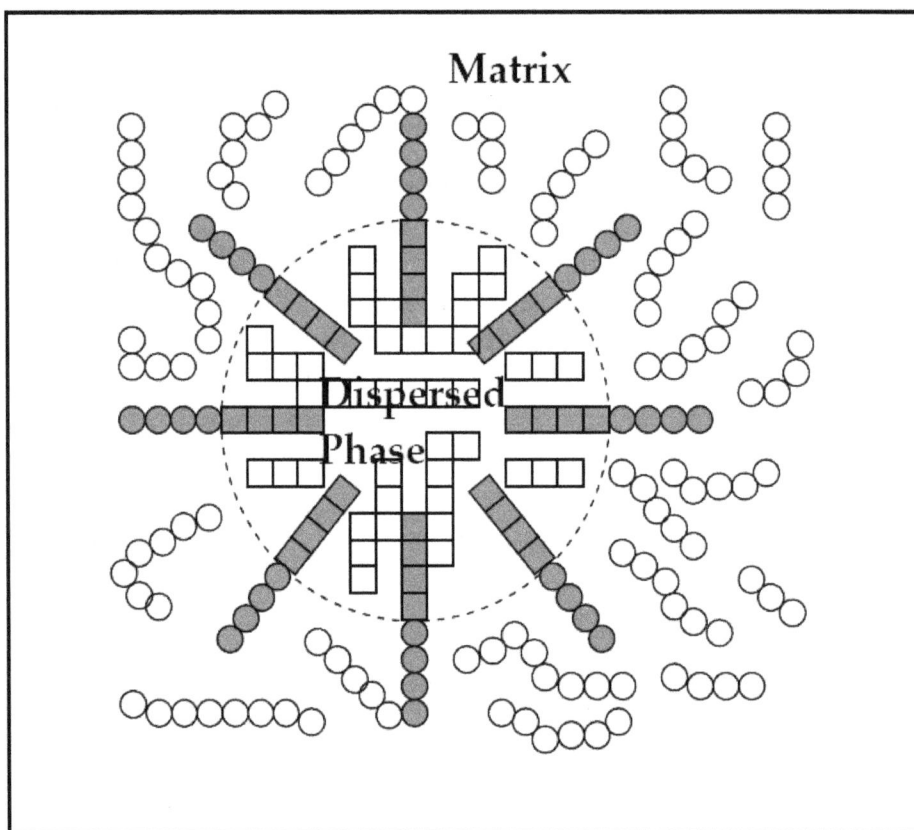

Figure 1 Illustration of the role of interfacial agent in compatibilizing blends

Starch-based blends by melt mixing

Starch-synthetic polymer blending has been studied as early as in 1973. The most often used synthetic polymer for blending with starch is polyethylene. The starch/polyethylene blends are used for agricultural mulch or food packaging. The uncompatibilized blends of starch and polyethylene show a coarse phase separation due to differences in polarity of starch (hydrophilic) and polyethylene (hydrophobic). The mechanical properties of these blends (tensile strength and elongation at break) decrease at higher starch content. Polyethylene-g-maleic anhydride (PE-g-MA) and polyethylene-g-glycidyl methacrylate (PE-g-GMA) have been used as reactive compatibilizer for starch/PE blends. PE-g-MA contains reactive anhydride sites, while PE-g-GMA posseses epoxide groups, which both can react in situ with the hydroxyl groups of starch. As a result a graft copolymer (PE-g-Starch), i.e. the compatibilizer, is formed upon mixing, thus improving the dispersion of starch in the PE matrix.

The mechanical properties of the blend are also improved. The tensile strength of the uncompatibilized blends is drastically reduced when starch content is increased, while the tensile strength of the compatibilized ones decreases only slightly with the starch content.

The use of conventional synthetic polymer such as PE for blending with starch will only result in a partially biodegradable material, since the conventional synthetic polymers are usually poorly or non-biodegradable. To obtain completely biodegradable products, synthetic biodegradable polymers have been applied, among which synthetic polyesters are considered very promising materials.

Examples of these biodegradable polyesters are poly-glycolide, poly-dioxanone, poly-lactides, and poly-lactones (such as poly-butyrolactone, poly-valerolactone, and poly-caprolactone). The ester bonds of these polymers are susceptible to attack by water and this leads to enhanced biodegradability. These biodegradable polyesters will finally decompose into non-toxic products. Some of these polyesters also have very good mechanical, thermal and water/gas permeability properties that are even comparable to bulk non-biodegradable polymers such as PE and PP, and EVOH.

Polycaprolactone (PCL) is well-known synthetic biodegradable polyester, which combines excellent biodegradability with acceptable mechanical properties. Studies on the blending of starch with PCL have been already described. As is the case of blends with PE, the uncompatibilized blends of starch with PCL give coarse phase separation and a reduction in the mechanical properties when the starch content is increased. The use of reactive compatibilizer precursors PCL-g-glycidyl methacrylate (PCL-g-GMA), PCL-g-pyromellitic anhydride, PCL-g-maleic anhydride (PCL-g-MA), dextran-g-PCL, and of a premade starch-g-PCL for starch/PCL blends resulted in a better dispersion between the phases and in turn in improved mechanical properties compared to the uncompatibilized blends. Despite these good results, systematic studies on the compatibilizer precursor synthesis as well as on the influence of the chain topology and chemical reactivity (for the in situ compatibilizer formation) have not yet been reported.

Starch-based blends by in situ polymerization

The synthesis of starch based graft copolymers by in situ polymerization represents not only an alternative route to melt blending (vide supra) for the production of novel biomaterials, but it can also provide an efficient synthetic methodology for the production of a compatibilizer to be used for melt blending of starch with a biopolymer.

In the past, starch-g-PCL has been synthesized using toxic materials such as isocyanates. Another approach consists of the Ring Opening Polymerization (ROP) of-caprolactone monomer in the presence of starch. In this reaction, the hydroxyl groups of starch are supposed to function as initiating sites. Previous studies showed that common ROP catalysts such as tin octoate or aluminium isopropoxide gave low (0-14%) grafting efficiencies (GE, defined as the percentage of grafted polyester to starch compared to the total amount of homopolymer and grafted polyester). Starch is a very hydrophilic material that always containsmoisture. The water in starch granules competes with the hydroxyl groups of starch in the initiation step of the polymerization reaction, leading to formation of PCL homopolymers rather that starch-g-PCL, thus resulting eventually in low GE values. Another possible cause for the low GE values is the heterogenous nature of the reaction. Starch is insoluble in the typical organic solvents used for ROP (such as toluene or THF), and the presence of liquid-solid reaction system leads to reduced reaction rates between starch and ε–caprolactone. The highest GE value (up to 90%) has been achieved when using triethylaluminium as catalyst, which however is extremely air - and water-sensitive, and difficult to handle since it releases ethane, a very flammable by-product, during the reaction. A new strategy for the in situ ROP of ε–caprolactone on starch with the use of common ROP catalyst is therefore highly desirable.

14
Chapter

VEGETABLE OIL-BASED BIO-POLYMERS: SYNTHESIS AND APPLICATIONS

INTRODUCTION

Health-related issues, stringent environmental protection policies, search for cost-effective and alternative materials, and the quest for renewability, sustainability, and high-performance materials for technical applications have led to intense research in the production of renewable thermosetting polymers from plant seed oils and shift in focus from the petrochemical based polymers. Bio-based polymers are gaining overwhelming interest and recognition worldwide due to the health, safety, and environmental concerns associated with the conventional synthetic polymers. These bio-based polymers are renewable, biodegradable, and environmentally-friendly. The examples of the bio-based polymers are thermosetting polymers from plant seeds such as soybean oil, sun flower oil, cashew nut oil and linseed oil rapeseed oil. There are also the thermoplastic counterparts such as poly-lactic acid PLA from maize/corn, poly-hydroxy buterate poly-capro lactone and so on.

Cost is a significant barrier to the development of renewable materials; however, their production has become a viable proposition as technologies have evolved. Inflation in the price of petroleum and an increasing awareness of the end-of-life disposal of the fossil-based plastics has also helped to establish the renewable materials.

Bio-based polymers can now be used as natural, sustainable alternatives to traditional petrochemical-derived materials such as phenol-formaldehyde, epoxy resin, unsaturated polyester resin, polyurethane, phenolic resin, and iso-cyanate resin in the manufacture of composite and in coating applications.

The polymer used in composite manufacturing is referred to as a matrix. The work of a matrix is to act as a binder and stress distributor. The stress should be transferred to the fiber that carries the load. Thermoset resins from plant seed oils are capable of replacing thermosetting resins from petrochemicals, particularly because of their positive environmental attributes. Unsaturated polyester resin, phenolic resin, formaldehyde, vinyl ester, polyurethane, and epoxy resin have been used extensively in composite manufacturing and some of them are in the coating industry. The plant oils cannot be used on their own in composite application unless they are suitably functionalized in order to add cross-linkable functionalities to the fatty acids.

Plant Oil Triglycerides

Triglycerides from plants, such as soy bean, palm, rapeseed or sun flower, can be utilised. The triglyceride compound must be isolated and purified, and also functionalised to obtain the requested reactivity. Various chemical modification reactions are possible; the most common goes via an epoxidation reaction. Therefore, not only would the use of plant oil based resins in liquid moulding resins reduce volatile organic compounds emissions, thereby reducing health and environmental risks, but it would also promote global sustainability.

Thermosetting polymers have several advantages over thermoplastics because they have higher service temperature, better stiffness, fatigue resistance, low creep, no stress relaxation, and relative ease of processing due to lower viscosity. Upon curing, thermosetting matrix forms a permanent 3-dimensional network that cannot be reprocessed. The cross-links tie all the polymer molecules together; even when heated, these molecules cannot flow past each other, or around each other. This is the reason they do not melt, and it is very difficult to break the molecules apart; thus, they cannot be remolded or reshaped once they are cross-linked. A major disadvantage of thermoset polymers is that they are difficult to recycle.

Plant oils are vegetable oils extracted from plant sources, as opposed to animal fats; they are quite abundant in nature. Soybean, linseed, castor, sunflower, rape seed, and palm oils are some examples of plant oils. Soybean oil, for example, has been used extensively in the food processing industry in salad dressings, sandwich spreads, margarine, and mayonnaise—and also in non-food applications such as inks, plasticizers, crayons, paints, and soy candles . Soybean oil (SBO) is the most readily available and one of the lowest-cost vegetable oils in the world. Linseed oil has been used mainly in non-food applications as an impregnator and varnish in wood finishing, as a pigment binder in oil paints, and as a plasticizer and hardener in putty. Inedible vegetable oils include processed linseed oil, tung oil, and castor oil, and they are widely used in lubricants, paints, coatings cosmetics, and pharmaceuticals. In general it would be right to say that plant oils are renewable raw materials for a wide variety of industrial products, including coatings, inks, plasticizers, lubricants and paints. Soybean oil is the most abundant of all plant oils, and it is mainly grown in the USA. Plant oils are triglycerides and contain various fatty acids such as linoleic, linolenic, oleic, palmitic, and stearic acid.

These fatty acids differ in chain length, composition, distribution, and location. Some are saturated and some are unsaturated, which results in differences in the physical and chemical properties of the oil. Plant oils are also relatively cheap.

FATTY ACIDS OF VEGETABLE OIL-BASED POLYMERS

Recently, polymers developed from renewable resources have attracted much attention due to their environmental and economic advantages. The growing environmental awareness and new rules and regulations are compelling the industries to seek more ecologically friendly materials for their products. Commercial markets for biodegradable and biobased polymers are expected to increase substantially in the coming years. However, the physical and chemical properties of conventional soybean oil limit its use for many industrial applications. Soybean resins are based on triglycerides, which are the major component of plant and animal oils. Triglycerides are composed of three fatty acid chains joined by a glycerol center.

The triglyceride molecule of typical plant seed oil. The fatty acids have various levels of saturation and unsaturation, and it is the unsaturated part that is functionalized to give the desired matrix which is used in composite manufacture. The liquid resin obtained after functionalization of the triglycerides with various chemical groups can also go through free radical polymerization reactions for complete curing. Many active sites from the triglycerides, such as double bonds, allylic carbons, and ester groups, can be used to introduce polymerizable groups. The chain lengths of the fatty acids in naturally occurring triglycerides can be of varying lengths; 16, 18, and 20 carbon atoms are the most common.

Triglycerides in plant oils typically contain 10 or more different fatty acids. Fatty acids can be saturated, unsaturated, isolated, or conjugated. Some of the unsaturated fatty acids can either be mono-unsaturated (containing 1 carbon-carbon double bond) or polyunsaturated (containing more than two carbon-carbon double bonds). The fatty acid composition of various oils. The iodine values of unsaturated fatty acids and their triglycerides.

The iodine value is directly related to the level of unsaturation of the fatty acids, which means that the higher the iodine value the higher the level of unsaturation. In soybean oil and linseed oil, linoleic acid and linolenic acid (respectively) predominate in their triglycerides, and the more unsaturated a fatty acid is, the more susceptible it is to functionalization. This means that linseed oil will be easier to modify than soybean oil since the linolenic acid predominating in linseed oil has three carbon-carbon double bonds in its chain, and thus more reactive sites, whereas linoleic acid (which predominates in soybean oil) has two carbon-carbon double bonds in its chain. Triglycerides are divided further into three main groups according to their iodine values. When the iodine value is more than 130, it is said to be drying oil; with iodine values of between 90 and 130, it is termed

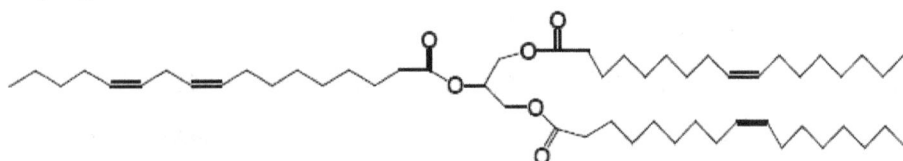

Figure 1. A triglyceride molecule, the major component of natural oils.

Table 1. Some fatty acids in natural oils

Myristic acid	$C_{14}H_{28}O_2$	$CH_3(CH_2)_{12}COOH$
Palmitic acid	$C_{16}H_{32}O_2$	$CH_3(CH_2)_{14}COOH$
Palmitoleic acid	$C_{16}H_{30}O_2$	$CH_3(CH_2)_5CH=CH(CH_2)_7COOH$
Stearic acid	$C_{18}H_{36}O_2$	$CH_3(CH_2)_{16}COOH$
Oleic acid	$C_{18}H_{34}O_2$	$CH_3(CH_2)_7CH=CH(CH_2)_7COOH$
Linoleic acid	$C_{18}H_{32}O_2$	$CH_3(CH_2)_4CH=CH-CH_2-CH=CH(CH_2)_7-COOH$
Linolenic acid	$C_{18}H_{30}O_2$	$CH_3-CH_2-CH=CH-CH_2-CH=CH-CH_2-CH=CH(CH_2)_7COOH$
α-Eleostearic acid	$C_{18}H_{30}O_2$	$CH_3-(CH_2)_3-CH=CH-CH=CH-CH=CH(CH_2)_7-COOH$
Ricinoleic acid	$C_{18}H_{33}O_3$	$CH_3(CH_2)_4CH-CH-CH_2-CH=CH(CH_2)_7-COOH$ $\qquad\qquad\;$ OH
Vernolic acid	$C_{18}H_{32}O_3$	$CH_3(CH_2)_4CH-CH-CH_2-CH=CH(CH_2)_7-COOH$ $\qquad\qquad\;$ O
Licanic acid	$C_{18}H_{28}O_3$	$CH_3(CH_2)_3CH=CH-CH=CH-CH=CH(CH_2)_4C-(CH_2)_2COOH$ $\qquad\qquad\qquad\qquad\qquad\quad$ O

Table 2. Fatty acid composition of various oils

Fatty acid	Castor oil (%)	Linseed oil (%)	Oiticica oil (%)	Palm oil (%)	Rape seed oil (%)	Refined tall oil (%)	Soybean oil (%)	Sunflower oil (%)
Palmitic acid	1.5	5	6	39	4	4	12	6
Stearic acid	0.5	4	4	5	2	3	4	4
Oleic acid	5	22	8	45	56	46	24	42
Linoleic acid	4	17	8	9	26	35	53	47
Linolenic Acid	0.5	52	-	-	10	12	7	-
Ricinoleic Acid	87.5	-	-	-	-	-	-	-
Licanic acid	-	-	74	-	-	-	-	-
Others	-	-	-	2	2	-	-	-

Table 3. Iodine values of unsaturated fatty acids and their triglycerides

Fatty acid	Number of carbon atoms	Number of double bonds	Iodine value of acid	Iodine value of triglyceride
Palmitoleic acid	16	1	99.8	95.0
Oleic acid	18	1	89.9	86.0
Linoleic acid	18	2	181.0	173.2
Linolenic acid and	18	3	273.5	261.6
α-Eleostearic acid				
Ricinoleic acid	18	1	85.1	81.6
Licanic acid	18	3	261.0	258.1

Semi-drying oil; and with a value of less than 90, it is non-drying oil . The average degree of unsaturation is measured by the iodine value.

THE MODIFICATIONS OF VEGETABLE OILS BASED ON TRIGLYCERIDE ESTER

Monomers Based on Carbon-Carbon Double Bond Modifications in the Vegetable Oils

The carbon–carbon double bonds in the fatty acid chains of the vegetable oils can undergo various reactions to attach different polymerizable functionalities, such as acrylates, to increase the reactivity of the vegetable oils.

Acrylated epoxidized soybean oil (AESO), synthesized from the reaction of acrylic acid with epoxidized soybean oil has been extensively studied in polymers and composites.

Monomers Based on Triglyceride Ester Modifications in the Vegetable Oils

The modifications of the fatty acid carbon-carbon double bonds, the incorporation of more reactive carbon– carbon double bonds through chemical modifications of the triglyceride ester groups is another promising approach to more reactive monomers. Wool and co-workers have developed a series of vegetable oil monoglyceride maleates copolymerized with ST to give rigid thermoset polymers. Figure 3 shows the preparation and polymerization of soybean oil monoglyceride (SOMG) maleate half esters.

THE APPLICATION AREAS

The utilization of renewable resources in energy and material applications is receiving increasing attentions in both industrial and academic settings, due to concerns regarding environmental sustainability .

The use of vegetable oils as renewable raw materials for the synthesis of various monomers and polymeric materials is reviewed. Vegetable oils are generally considered to be the most important class of renewable resources, because of their ready availability and numerous applications. Recently, a variety of vegetable oil-based polymers have been prepared by free radical, cationic, olefin metathesis, and condensation polymerization. The polymers obtained display a wide range of thermo physical and mechanical properties from soft and flexible rubbers to hard and rigid plastics, which show promise as alternatives to petroleum-based plastics.

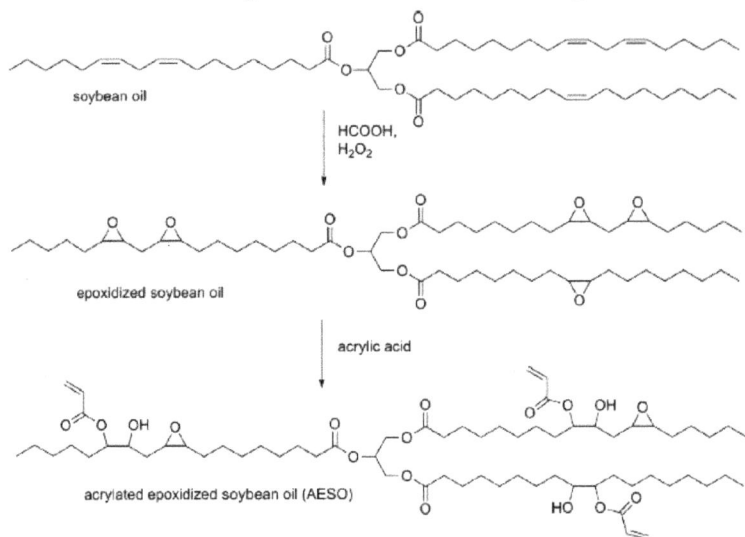

Figure 2. Synthesis of acrylated epoxidised soybean oil

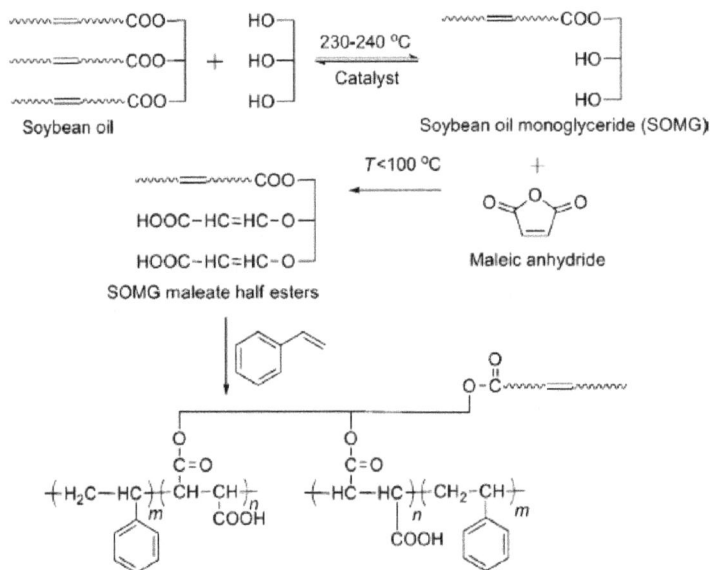

Figure 3. Synthesis and polymerization of soybean oil monoglyceride maleates.

Triglyceride based oil from renewable resources are now being used in paints, adhesives and composite industry. Their application as matrix in composite industry has opened new markets in construction, automobile, marine, military, sports and leisure. The tight environmental regulations have forced many companies to embrace the use of the renewable materials. The application areas of the vegetable oil-based polymers can be further explored.

CONCLUSION

Plant seed oils are renewable alternatives to petrochemicals, but they cannot be used in their raw form except they are suitably modified. The unsaturated carbon-carbon double bonds in the fatty acid must be made to react with other monomers because the aim of the modification is to attach cross linkable functionalities to the fatty acid in order to have cross-linked thermoset material after curing. After modification and subsequent curing of the polymer, the final product should be a rigid thermoset material which cannot be reprocessed or melted. The level of unsaturation of the various fatty acids determines the type of polymerization that should be carried out. The in-depth study and the understanding of the fatty acids of the plant seed oils are paramount for their modification and further application in high technical area.

15
Chapter

RENEWABLE BIO-POLYMERS FROM TUNG OIL AND NATURAL TERPENES

INTRODUCTION

The search for sustainable chemistry has led to the development of new synthetic routes to produce polymers from renewable resources. Lately, particular attention has been paid to biomass as a source of a wide range of biomaterials. Amongst many biomass derived materials, it is important to highlight the potential use of vegetable oils. The employment of these materials as platform chemical for polymer synthesis has several advantages, which include availability, inherent biodegradability, high purity and low toxicity. Vegetable oils or triglycerides are fatty acid esters of the triol glycerol. Basically, a fatty acid consists of a hydrophobic hydrocarbon chain with a hydrophilic group at one end. The degree of unsaturation varies according to the composition of fatty acids presenting double bonds. This value, one of the most important parameters in the chemistry of fatty acids and triglycerides, can be calculated through the iodine value, a structural index which has been related to various physical and chemical properties and has served as a quality control method in hydrogenation reactions.

Terpenes, natural monomers having carbon skeletons of isoprene units, are found in many essential oils and represent a versatile chemical feedstock. Even though the use of terpenes in chemical synthesis is vast and well supported by a large number of publications, their applications in polymer science are scarce and far between, usually limited to attempts to polymerize bicyclic monoterpenes, such as α and β pinene by Lewis acid and Ziegler-Natta catalysis. As stated by Norström, terpenes do not tend to polymerize, and their lack of reactivity can be attributed to both steric hindrance and low stabilization energy between the monomer and the chemical species in the transition state. However, since terpenes exhibit carbon-carbon double bonds amenable to reactions, several works have been published concerning the polymerization of terpenes with vinyl monomers in order to develop biodegradable polymeric materials.

POLYMER PREPARATION VIA CATIONIC POLYMERIZATION

The polymeric materials were prepared by cationic polymerization initiated by modified BFE. The desired amount of the comonomer was measured out and added to the tung oil. The mixture was vigorously stirred followed by the addition, drop by drop, of the appropriate amount of the modified initiator. The modified catalyst was prepared by dissolving the boron trifluoride diethyl etherate and 1, 4-butanediol in the desired amount of THF. The reaction mixture was placed into an aluminum mold and heated for a given time, usually 12 hrs. at room temperature, followed by 12 hrs. at 60°C and finally for 12 hrs. at 110°C.

Cationic Polymerization of Vegetable Oils

Early in this research, several attempts to develop oil-based materials by cationic polymerization; employing different types of vegetable oils, such as linseed oil, tung oil, soybean oil, corn oil, etc. were carried out. Keeping in mind the preparation of completely bio-based polymers as the major goal, the triglycerides were copolymerized with terpenes, natural monomers containing alkenyl functional groups amenable to react.

Fruitful reactions leading to useful materials were achieved when tung oil, whose major component is a fatty acid containing three conjugated unsaturations, was copolymerized with limonene and myrcene as reactive comonomers.

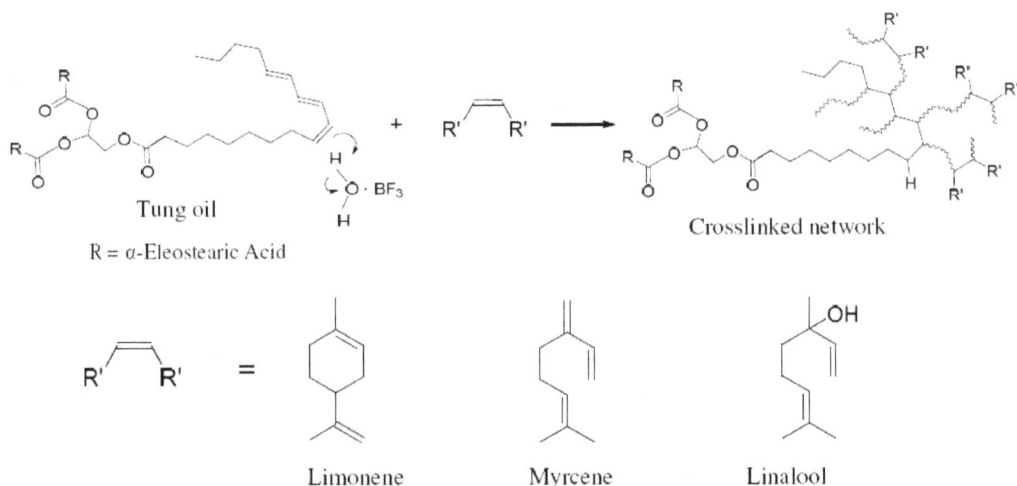

Figure 1 Cationic polymerization of tung oil with limonene, myrcene and linalool

As pointed by Liu and Erham, the polymerization of 1, 2 disubstituted ethylenes containing small side groups easily occurs. Nevertheless, when bulky groups are attached to the double bonds, the polymerization either does not occur or leads to low molecular weight products after long reaction times. The steric hindrance plays a preponderant role during the propagation of the growing chains. It introduces a high energy barrier between the monomer and the growing species, allowing a very limited amount of polymer to be produced. Since fatty acids are 1, 2 disubstituted ethylene monomers bearing bulky groups, their polymerization is considered energetically unfavorable.

Unlike the other tested oils, tung oil was found to be highly reactive towards cationic and free radical polymerizations. Its particular reactivity can be associated to its capacity to stabilize, by resonance, the intermediate carbocation generated on the propagation step of the cationic mechanism. In regards to the use of terpenes as comonomers, successful polymerizations were carried out when using limonene and myrcene. The polymerization of tung oil with linalool led to highly porous materials. The gassing generated during curing could be attributed to side reactions between the linalool and the boron trifluoride ethereate. All the copolymers appeared as dark-brown rubbery materials at room temperature. For all the studied comonomers, suitable materials for analysis were not obtained for terpene proportions above 60 wt%.

Summarizes the properties of the tung oil-limonene copolymers. As shown, the storage modulus, the crosslinking density and the Young`s modulus decreased when the tung oil concentration increased. On the other hand, Tg decreased almost linearly with the increase in the limonene content. Pycnometry showed variations in the density of the copolymers for percentages of limonene higher than 30 wt%.

Table 1 Properties of the tung oil-limonene copolymers

Limonene Proportion (wt %)	Storage Modulus (MPa)	Tg (°C)	Crosslinking Density, n (mol/m³)	Density, (g/cm³)	Young's Modulus (MPa)
0	32.5	18.28	4118	1.0036 ± 0.0027	33.80 ± 2.92
10	17.8	13.89	2287	1.0097 ± 0.0020	14.95 ± 1.90
20	11.4	10.53	1480	1.0068 ± 0.0132	10.40 ± 0.26
30	4.5	5.58	594	1.0080 ± 0.0252	6.34 ± 1.62
40	2.7	4.33	357	0.9839 ± 0.0036	3.38 ± 0.31
50	0.8	0.17	112	0.9764 ± 0.0151	3.26 ± 0.07
60	0.5	-3.83	65	0.9414 ± 0.0081	2.22 ± 0.12

Fatty acid chains can contribute to two different effects, either plasticizing the network because of the more flexible structure or increasing the rigidity due to the large number of unsaturations per molecule. Both factors can contribute to the properties of the copolymers. In the case of tung oil-limonene materials, a decrease in both the number of crosslinking points and the glass transition temperature of the networks was observed. As demonstrated by Norstrom, limonene does not tend to homopolymerize (under the experimental conditions reported in Norstrom`s work, polymerizing limonene units through cationic polymerization mainly yielded to oligomers that consisted of trimers and tetramers). However, even though it is been clearly stated that formation of large molecules (homopolymerization) out of this terpene not likely to occur, several works have proved that limonene is able to react with vinyl monomers leading to copolymers in which terpene units have been incorporated into the vinyl resin. Due to the inability of limonene to form long chains (even at high weight proportions), here it is suggested that the decrease in the number of crosslinking points arises as a result of limonene units that have been incorporated into the fatty acid chains, lowering their ability to form an oil-based cross-linked network. The decrease in the glass transition of the networks at high limonene contents, in which crosslinking is very low, is driven by the glass transition of the limonene oligomers (small linear segments) formed during heating of the polymeric matrices.

DMA of the tung oil-myrcene copolymers revealed a similar behavior to that shown by the tung oil-limonene copolymers. When the temperature increased, the storage modulus exhibited a drop over a wide temperature range, followed by a modulus plateau at high temperatures, evidencing the presence of a stable cross-linked network. Also, the modulus decreased as the myrcene proportion in the mixtures increased. The transition between the glassy to the rubbery state for these copolymers was wide and began below room temperature. Likewise, for all the copolymers, no phase separation was observed. The results showed that the tan δ peak increased in height and become narrower as the myrcene content in the matrix increased. As in the case of tung oil-limonene copolymers, this phenomenon could be attributed to the presence of less heterogeneous networks. Unlike the tung oil-limonene materials, the maximum in the loss factor shifted to higher temperatures when increasing the myrcene proportion above 10 wt%, indicating that the glass transition temperature for these copolymers increased as the comonomer proportion increased. Summarizes some of the properties of the tung oil-myrcene copolymers. As shown, the glass transition temperature increased with the content of myrcene for percentages higher than 10 wt%. Likewise, the increase in the comonomer proportion decreased the amount of crosslinking points in the copolymers. The storage modulus, along with the Young's modulus, decreased with the content of myrcene. The pycnometry test did not show appreciable changes in the density with the change in the monomer proportion.

Myrcene is a natural conjugated diene that has been researched as a monomer for the preparation of 100 wt% biosource elastomeric materials. Polymyrcene has been prepared by several methods, including anionic polymerization with sec-BuLi and more recently, was homopolymerized by a free radical process in aqueous media.

Table 2 Properties of the tung oil-myrcene copolymers

Myrcene Proportion (wt %)	Storage Modulus (MPa)	Tg (°C)	Crosslinking Density, n (mol/m³)	Density, (g/cm³)	Young's Modulus (MPa)
0	32.5	18.28	4118	1.0136 ± 0.0057	33.80 ± 2.92
10	22.5	13.38	2905	1.0097 ± 0.0028	9.59 ± 1.72
20	14.6	15.31	1867	1.0098 ± 0.0082	5.36 ± 0.44
30	7.3	17.71	925	1.0102 ± 0.0152	3.22 ± 0.15
40	3.4	22.32	420	1.0134 ± 0.0360	1.31 ± 0.06
50	1.3	25.44	156	1.0092 ± 0.0101	059 ± 0.01
60	0.3	29.82	33	1.01.9 ± 0.0181	0.21 ± 0.03

Myrcene yields polymerswith different structures, due to the three possible electrophilic additions to the conjugated diene placed in the molecule; being the 1, 4 addition the predominant one (77-85%) and 3,4 addition in lesser degree (15-23%) . Unlike other common 1, 3-dienes (such as butadiene or isoprene), myrcene provides polymers with structural units bearing an additional double bond in its alkyl tail, which is very suitable for crosslinking.

Myrcene is a natural conjugated diene that has been researched as a monomer for the preparation of 100 wt% biosource elastomeric materials. Polymyrcene has been prepared by several methods, including anionic polymerization with sec-BuLi and more recently, was homopolymerized by a free radical process in aqueous media. Myrcene yields polymers with different structures, due to the three possible electrophilic additions to the conjugated diene placed in the molecule; being the 1, 4 addition the predominant one (77-85%) and 3, 4 addition in lesser degree (15-23%). Unlike other common 1, 3-dienes (such as butadiene or isoprene), myrcene provides polymers with structural units bearing an additional double bond in its alkyl tail, which is very suitable for crosslinking.

When polymerizing myrcene and tung oil in the presence of BFE, the myrcene, contrarily to the limonene, is not only able to react with the oil, but it is also capable of homopolymerizing and creating linear myrcene-myrcene sequences. When increasing the amount of comonomer, the diluting effect of the myrcene sequences immersed into the resin becomes predominant, and hence, the crosslinking density decreases, consequently leading to a decrease in the storage modulus. The increase in the glass transition of the networks when increasing the myrcene content is attributed to the presence of these linear myrcene-myrcene sequences formed during heating. As mentioned above, these linear polymyrcene sequences contain structural repeating units bearing very short dangling chains having an extra double bond, which can be prompt to react under the experimental conditions. At high myrcene concentrations, these additional unsaturations can react and generate polymyrcene cross-linked networks. The bonding of these polymyrcene linear segments (embedded in the matrix) through crosslinking, stiffs and strengths the system and therefore; limits the segmental motion of the polymer chains, driving the glass transition temperature to higher values whilst increasing the terpene content.

In order to compare the results obtained in this research with previous works dealing with copolymerization of vegetable oils and synthetic vinyl monomers, tung oil was copolymerized with styrene in the presence of boron trifluoride as an initiator, at room temperature. Summarizes the properties of the tung oil-styrene copolymers. The glass transition temperatures (Tg) increased almost linearly with the content of styrene for percentages higher than 10 wt%.

The increase in tung oil concentration increased the amount of crosslinking points in the copolymers. The storage modulus, along with the Young`s modulus, decreased with the content of styrene. The pycnometry test also did not show any appreciable change in the density with the variation in the monomer proportion.

Table 3 Properties of the tung oil-styrene copolymers

Styrene Proportion (wt %)	Storage Modulus (MPa)	Tg (°C)	Crosslinking Density, n (mol/m³)	Density, (g/cm³)	Young's Modulus (MPa)
0	32.5	18.28	4117	1.0136 ± 0.0057	33.80 ± 2.92
10	14.7	10.00	1913	1.0129 ± 0.0087	17.10 ± 2.07
20	10.7	12.04	1383	1.0158 ± 0.003	10.17 ± 0.76
30	8.7	15.91	1109	1.0205 ± 0.0121	8.98 ± 0.35
40	6.8	20.15	858	1.0238 ± 0.0028	7.30 ± 0.67
50	4.7	27.32	584	1.0292 ± 0.0156	6.29 ± 0.23
60	2.9	30.47	354	0.9653 ± 0.0178	5.38 ± 0.39

Just like in the cases of tung oil-limonene and tung oil-myrcene, the modulus was found to be strongly dependent on the comonomer proportion. It exhibited a drop over a wide temperature range, followed by a modulus plateau at high temperatures, evidencing the presence of a stable cross-linked network. All the copolymers showed only one tan δ peak denoting, like in the other cases, the homogeneity of the materials. The height of the tan δ peak increased as the styrene content in the matrix increased. The maximum in the loss factor shifted to higher temperatures when increasing the comonomer weight proportion, denoting an increment in the glass transition temperature. As myrcene, the styrene is not only able to react with the oil, but also, it is capable of homopolymerizing and generating linear styrene-styrene sequences incorporated between the oil matrixes.

As increasing the amount of comonomer, the diluting effect of these linear segments immersed into the resin became predominant; leading to a decrease in the crosslinking density (consequently conducing to a reduction in the storage modulus). The increase in the Tg is also attributed to the presence of these stiff and strong linear Polystyrene segments embedded in the resins.

CONCLUSIONS

Thermosets were prepared through copolymerization of tung oil with limonene, myrcene and styrene. The Tg was found to be a function of both the type and comonomer weight content. The Young's modulus and the experimental crosslinking densities decreased as comonomer weight content increased. Based on the thermo-mechanical evaluation, these materials behave as elastomers at room temperature. The Fox and Loshaek model showed a relatively good prediction of the Tg values for tung oil-limonene copolymers. For myrcene, Tg values matched those predicted by the Fox equation. Even though triglyceride-based polymers are complex materials, some of the thermo-mechanical properties were found to be predictable by simple mathematical approaches and models developed for polyvinyl resins.

16
Chapter

TRIGLYCERIDES AND PHENOLIC COMPOUNDS OF BIO-BASED THERMOSETTING EPOXY RESINS

INTRODUCTION

In recent years, there has been an increasing interest for the development and application of biomass derived products to address issues related to the depletion of non-renewable resources.

Currently, the chemical industry is pursuing to substitute a growing part of fossil feedstocks with renewable carbon. This trend is not only driven by environmental concerns, but it is also fueled by the increase in the general demand for green products from costumers; as well as efforts from government agencies providing funding to increase the present-day knowledge and technology to process and transforming biomass into high-value chemicals . In order to fulfill the needs of the current industry, the polymer industry also demands a renewable chemical platform; and the main challenge lies in developing materials with properties matching or even improving those of resins in current use. In the particular case of thermosetting materials, which have played an important role in the modern civilization, these properties include high modulus, strength, durability, as well as the thermal and chemical resistances provided by the high crosslinking density assembling their macromolecular structure.

In non-food applications, the most widely applied renewable resources include plant oils (only castor and linseed oil are exclusively used of industrial applications), polysaccharides (mainly starch and cellulose), and proteins. Ronda et al. have considered three main different approaches to convert oils and their derivatives into useful polymeric materials; the direct polymerization of the double bonds, the chemical modification and posterior polymerization, and the polymerization of monomers synthesized using oil-derived platform chemicals. It is clear however, that this scheme can be extended not only to oils, but also to any other type of biomass-derived compounds, leading to three different areas to generate macromolecular materials, including direct use, chemical functionalization, and biorefinery.

SYNTHESIS OF EPOXIDIZED TRIGLYCERIDES

Linseed oil (LO) was epoxidized using a peracetic acid generated in situ from reaction between acetic acid and hydrogen peroxide, with ion-exchange resin catalyst, at 60°C in toluene, during 8 h. A solution of linseed oil (20.1 g, 67.0 mmol) and acetic acid (1.0 mL, 26.2 mmol) in toluene, using Amberite as an ion exchange catalyst, was stirred at 35°C, and H2O2 30% w/v was added dropwise. After H2O2 addition was complete, the temperature was raised to 60°C and the flask contents were stirred vigorously for 8 h. Once the reaction time was completed the mixture was first dissolved in ethyl acetate. The Amberlite was filtered off, and the filtrate was poured into a separation funnel, and washed with brine until the pH was neutral. The oil phase was further dried above anhydrous sodium sulfate and then filtered. Finally, the solvent was removed using a rotary evaporator.

Synthesis of polyol

Polyol was synthesized from epoxidized linseed oil, with an oxirane oxygen content of 0.653 mol/100g, through a ring opening reaction with boiling methanol in the presence of tetrafluoroboric acid as acid catalyst following the procedure described by Dai et al. The molar ratio of epoxy groups to methanol was 1:11, and the concentration of the catalyst was 1 wt% of the total weight of oil and methanol. Methanol and catalyst were placed into the reactor and heated to approx 60°C using an oil bath. Once the temperature was reached, the epoxidized linseed oil was added dropwise. The reaction mixture was kept for 1 hour. After cooling to room temperature, the acid was neutralized with sodium bicarbonate aqueous solution (approx. 30 wt%) and the remaining solvent was removed on a rotatory evaporator under low vacuum.

Functionalization of phenolic compounds with epoxy groups. Synthesis of triglycidylated ether of α-resorcylic acid.

The glycidylation of bio-oil was conducted following the method and conditions established by. The described process leads to the synthesis of epoxyalkylaryl ethers by the reaction between compounds containing acid hydroxyl groups and haloepoxyalkanes in the presence of a strong alkali. A reactor was loaded with the desired amount of bio-oil (containing a certain amount of moles of -OH / gram of oil) and epichlorohydrin (5 moles epichlorohydrin / mole - OH). The mixture was heated at 100°C and benzyltriethylammonium chloride (0.05 moles BnEt3NCl / mole -OH) was added. After 1 h, the resulting suspension was cooled down to 30°C, and an aqueous solution of NaOH 20 wt. % (2 moles NaOH / mole –OH and also containing 0.05 moles of BnEt3NCl / mole –OH) was added dropwise. Once the addition of the alkaline aqueous solution was completed, the mixture was vigorously stirred for 90 min at 30°C, and then the reaction was stopped. After that, the mixture was poured into an extraction funnel and the organic layer was separated. Through numerous liquid-liquid extractions with deionized water, the organic phase was abundantly washed until neutral pH was reached, and then vacuumed concentrated. The crude resulting product was purified by silica gel chromatography using petroleum ether/ethyl acetate solvent system.

Synthesis of bio-based epoxy resins

In order to generate a polymeric material, epoxy monomers can undergo step-growth polymerization, using a crosslinking agent (such as anhydrides, amines, etc.) and chain homopolymerization in the presence of both, Lewis acids and bases, such as tertiary amines (known as anionic homopolymerization). As a first approach, epoxidized linseed oil was cured with a commercial hardener (SC-15 part B by Applied Poleramic Inc. composed of cycloaliphatic amine and polyoxyl-alkylamine) in different weight ratios. The mixture was cured for 6 hrs. at 60°C, 6 hrs. at 80°C, 2 hrs. at 120°C and finally 1 hr. at 160°C. In order to minimize the use of commercial agents in the formulation, epoxidized linseed oil and epoxidized α-resorcylic acid were mixed in different weight proportions and cured by means of a chain Homopolymerization. 4-dimethylaminopyridine or DMPA (a Lewis base) was added to the mixtures in a ratio of 0.08 mol/mol epoxy groups. The mixtures were then poured in an aluminum mold and cured at 120°C for 12 hrs and post cured for 2 hrs at 160°C.

Preparation of the linseed oil based polyurethane

The PUR plastic sheets were prepared following the procedure described by Kong and Narine by reacting the resulting polyol with aromatic TDA. As a first approach, a 1.0/1.1 molar ratios of the -OH group to the isocyanate (NCO) group (Mratio), was chosen for the formulation.

The desired OH/NCO molar ratio satisfies the Equation 3.1:

$$\text{Mratio} = \frac{\text{W polyol} / \text{EW polyol}}{(\text{W polyurethane} - \text{W polyol}) / \text{EW isocyanate}} \quad (3.1)$$

Here Wpolyol is the weight of the polyol, EWpolyol is the equivalent weight of polyol, WPU is thetotal weight of polyurethane to produce, and EWisocyanate is the equivalent weight of the isocyanate. The equivalent weights of polyol were determined using Equation 3.2:

$$\text{EW polyol} = \frac{56110}{-\text{OH number}} \quad (3.2)$$

The weight of the polyol and isocyanate were calculated using the above calculated equivalent weight. Suitable amounts of polyol and MDI were weighed in a plastic container and stirred slowly for 2 min. The mixture was cast directly into a metallic mold previously greased with silicone release agent and placed in an oven at 80°C for polymerization. The samples were postcured at 120°C for 12 h.

RESULTS AND DISCUSSION

Synthesis of epoxidized triglycerides

Previous investigations revealed that at room temperature, all the polymers are in the transition from the glassy state to the rubbery state, when using direct polymerization as an approach to obtained triglycerides-based polymeric materials (without any chemical modification of the starting materials). Since one of the main goals of this research is to develop and synthetize polymeric materials from renewable resources with high performance and outstanding properties (such as high modulus, high strength, high glass transition, chemical resistance, etc.); the driving force behind this project is to explore and study the chemical functionalization of plant oils. The modification of the triglyceride double bonds to introduce readily polymerizable functional groups has been used as a common strategy for obtaining high performance polymeric materials. The unsaturated fatty acid content of commercially available vegetable oils varies depending on the fatty acid composition. It determines the degree of unsaturation of the vegetable oil, and consequently, its reactivity. Also, the degree of unsaturation defines the content of oxirane rings, when undertaking epoxidation reactions. Table 1 shows the degree of unsaturation, along with the maximum theoretical oxirane content after epoxidation of some common oils.

Table 1 Properties of some common oils

Oil	Degree Unsaturation	Oxirane Content (gr/100 gr sample)
Linseed	6.25	10.2
Soabean	4.61	7.95
Canola	3.91	6.59

As seen in Table 1, linseed oil exhibits the highest degree of unsaturation, and thus, maximum theoretical oxirane content. The linseed oil based epoxy resin (ELO) was synthesized by epoxidation of the raw oil, using peracetic acid generated in situ (Figure 1).

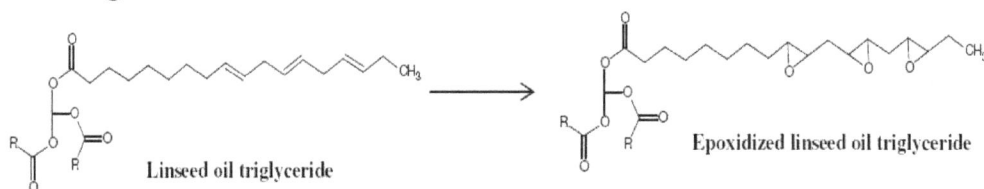

Figure 1 Epoxidation of linseed oil triglycerides

It was found that the optimum molar ratio of C=C/CH3COOH/H2O2 for obtaining the maximum conversion of C=C into epoxy rings, minimizing side reactions, was 1/0.5/2 at 60°C during 8 hrs. The iodine value and the oxirane value are important characteristics in the epoxidation of oils. The iodine value specifies the unsaturation degree of the oils, whereas the oxirane value determines epoxy content of the modified oil. Once the reaction was completed and the product isolated and purified, both parameters were quantified through titrations. According to the experimental results, the conversion of the double bonds to epoxy groups during the epoxidation reaction was found to be around 90%. From iodine values before and after epoxidation, it was found that around 98% of the total double bonds were removed during reaction. This indicates that approximately 8% of epoxy groups were lost due to side reactions. Titration with HBr in acetic acid revealed that this material has an average molar epoxy content of $(5.8 \pm 0.7) \times 10^{-3}$ moles per gram of resin. FTIR analysis supported the grafting process of epoxy functions onto the fatty acids chains of the triglyceride. The FTIR spectrum of triglyceride is characterized by the absorption band of -C=C- at 3100 cm-1 and the carbonyl group at 1750 cm-1.

After epoxidation, the peak at 3100 cm-1 is removed due to the transformation of the double bonds into oxirane groups; which is also evident by the presence of the absorption peaks at approximately 820 cm-1.

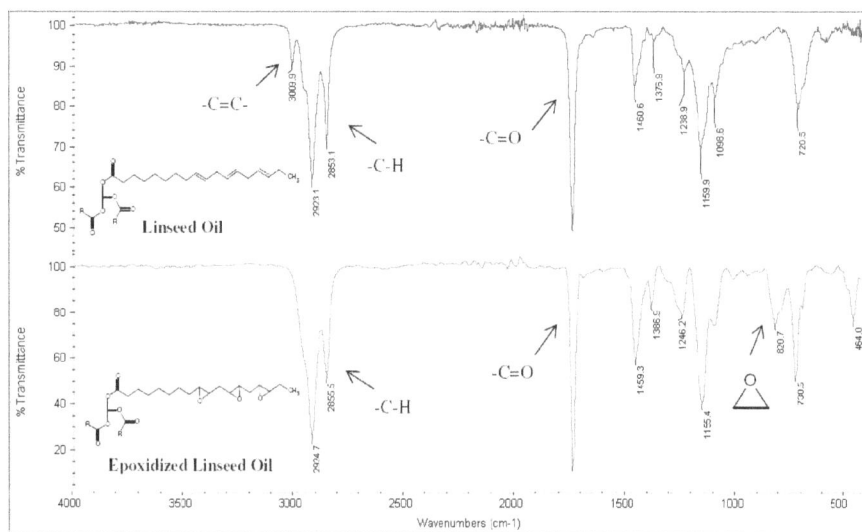

Figure 2 FTIR spectrum of epoxidized linseed oil.

SYNTHESIS OF LINSEED OIL BASED POLYOL

The linseed oil based polyol was synthesized by epoxidation of the raw oil, and its subsequent ring opening reaction with refluxing methanol using tetrafluoroboric acid as acid catalyst.

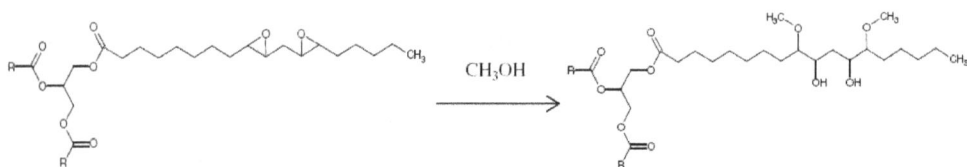

Figure 3 Methanolysis of epoxidized linseed oil

During the methanolysis reaction, the acid media promotes the oxirane ring opening in order to generate a hydroxyl functional groups attached to the fatty acid chains. In the first part of the mechanism, the acid reacts with the epoxide to produce a protonated epoxide, a highly electrophilic intermediate susceptible to the nucleophilic attack of the methanol molecules. Once the reaction is completed and the product isolated and purified, the number of –OH groups generated during reaction can be quantified through titrations. Table 2 summarizes the properties of the obtained polyol.

Table 2 Properties of the resulting linseed oil based polyol

Parameters	Hydoxyl Value HV (mg KOH/g) (*)	Conversion (%)	Theoretical Molecular Weight MW (g/mol)	Functionality[a] (*)
Values	230 (370)	75.0%	1072	4.2 (5.8)

a. Functionality = [(MW x HV)/56110] * Parenthesis indicates theoretical values

Using the composition of the linseed oil, the theoretical molecular weight and the theoretical hydroxyl number were calculated. Both parameters were based on the assumption that no dimmers or timers were formed, and that one hydroxyl group was formed per double bond. As seen, the conversion from epoxy groups to hydroxyl groups reached approx. 75.0% of yield. Just like in the case of the epoxidation reaction, there is side reactions involved in the ring opening reaction.

During the hydroxylation stage, the newly formed –OH groups, attached to the fatty acid chains, might attack the protonated epoxy ring leading to the formation of dimers and trimmers, which lowers the reaction yield. Another important parameter of a polyol, obtained from the hydroxyl value, is the functionality; which allows calculating the correct stoichiometric ratios when mixing polyols with multifunctional urethanes in order to produce polyurethanes. The functionality of the resulting polyol is lower than the number of double bonds per molecule of the starting oil, because a part of the functional groups was lost in the side reactions during the preparation of the polyol .

Besides analytical quantification, the chemical transformation of carbon-carbon double bonds into hydroxyl groups was also studied and traced through FTIR. As seen in Figure 4, the absorption bands of epoxidized triglycerides appeared within 1720-1700 cm-1 for C=O, at 825 and 845 cm-1 for epoxy groups, and the absence of the C=C stretching band at approximately 3070 cm-1. The final polyol product showed a broad peak within 3500-3200 cm-1, indicating the presence of –OH groups and the disappearance of the epoxy bands at 825 and 845 cm-1.

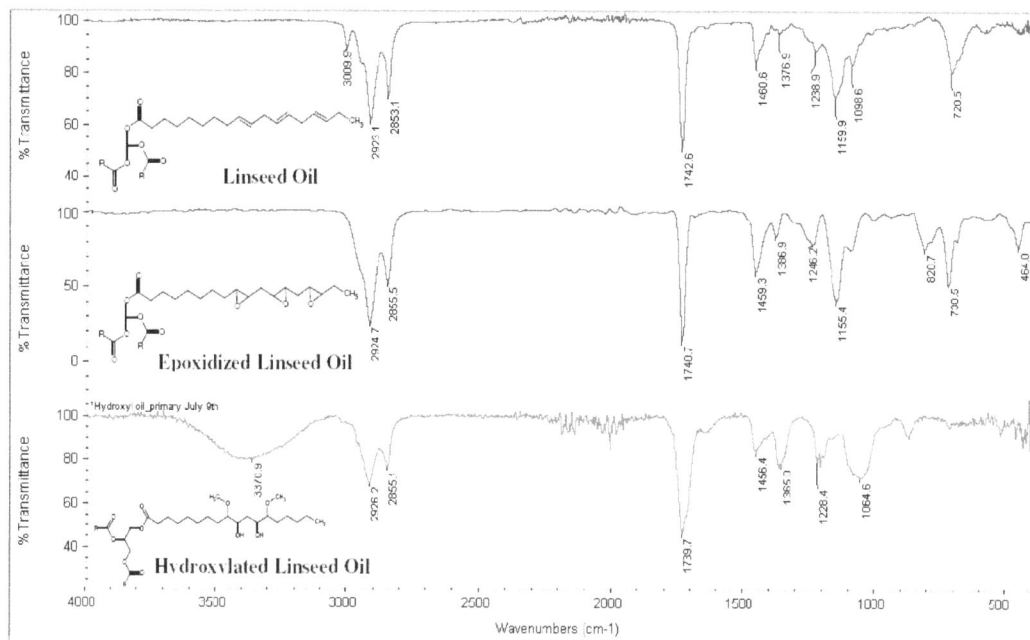

Figure 4 FTIR spectrum of hydroxylated linseed oil.

FUNCTIONALIZATION OF PHENOLIC COMPOUNDS WITH EPOXY GROUPS.

Synthesis of triglycidylated ether of α-resorcylic acid

Regarding the coupling of epoxy groups to the α-resorcylic acid, it was found that the treatment of this compound with epichlorohydrin in alkaline aqueous medium, in the presence of benzyltriethylammonium chloride as a phase transfer catalyst, allowed the total glycidylation of this phenolic acid to generate the triglycidylated derivative of α-resorcylic acid or TRA (also referred to as ERA, a short version of epoxidized α-resorcylic acid).

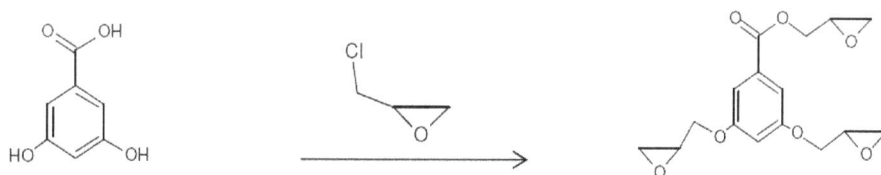

α- Resorcylic Acid   Triglycidylated ether of α-Resorcylic Acid (TRA)

Figure 5 Glycidylation reaction of α-resorcylic acid with epichlorohydrin.

The reaction yield, after purification through silica gel chromatography, was around 80 %. The remaining 20 % probably correspond to mono and di glycidilated products of α-resorcylic acid, as well as, unreacted material. Because phenols are more acidic than alcohols, they can be converted to phenolates through the use of a strong base. These intermediates are strong nucleophiles and can react with primary alkyl halides to form ethers. According to the literature, once the phenolate anions are formed, they can react with epichlorohydrin following two competitive mechanisms: one-step nucleophilic substitution (mechanism SN2) with cleavage of the C–Cl bond and a two-step mechanism based on ring opening of epichlorohydrin followed by intramolecular cyclization (through SN1 mechanism), with the release of chloride, to reform the epoxy ring. Titration with HBr showed that resulting mixture after modification has an average molar epoxy content of $(9.0 \pm 0.5) \times 10^{-3}$ moles per gram of resin, which corresponds to an epoxy equivalent weight = 111 gr/eq. The presence of epoxy groups was confirmed by FTIR.

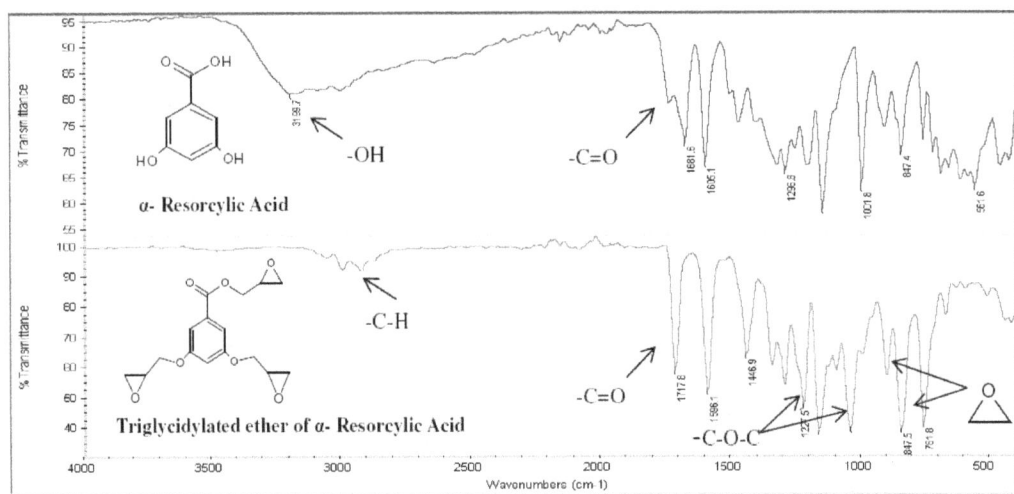

Figure 6 FTIR spectra of α-resorcylic acid and epoxidized α-resorcylic acid

The α-resorcylic acid structure was characterized by the presence of a strong IR absorption band at 3200 cm-1. The carbonyl exhibited a strong peak at 1750 cm-1. The spectrum of the epoxidized α-resorcylic acid presented similar absorption bands to those exhibited by its non-modified counterpart. The most remarkable difference between them was the disappearance of the weak peak at 3200 cm-1 (attributed to the presence of hydroxyl groups). The lack of this absorption peak in the FTIR spectrum of the epoxidized α-resorcylic acid is related to the removal of the -OH functional groups due to the glycidylation reaction. It is also evident the presence of absorption bands at 1227.5 cm-1 and 1050 cm-1 due to the presence of -C-O-C- groups, formed during the reaction, as well as the characteristic bands for oxirane rings at 847.5 cm-1 and 910 cm-1.

Synthesis of diglycidylated ether of α-resorcinol

With the aim of correlating the mechanical properties of the resulting resins with the number of functionalities of the epoxy monomers, resorcinol (containing two phenolic –OH groups) was also modified with epichlorohydrin. The reaction yield, after purification through silica gel chromatography, was around 85%. The remaining 15% probably corresponded to monoglycidylated products and unreacted resorcinol. Titration with HBr showed that the resulting mixture after chemical modification has an average molar epoxy content of $(8.7 \pm 0.5) \times 10^{-3}$ moles per gram of resin, resulting in an epoxy equivalent weight = 115 g/eq. The presence of epoxy groups was also confirmed by FTIR (Figure 8). The lack of the absorption peak at around 3200 cm-1 evidenced the transformation of phenolic hydroxyl groups into glycidyl moieties. Bands for oxirane rings at 847.5 cm-1 and 910 cm-1 confirmed the presence of epoxy groups.

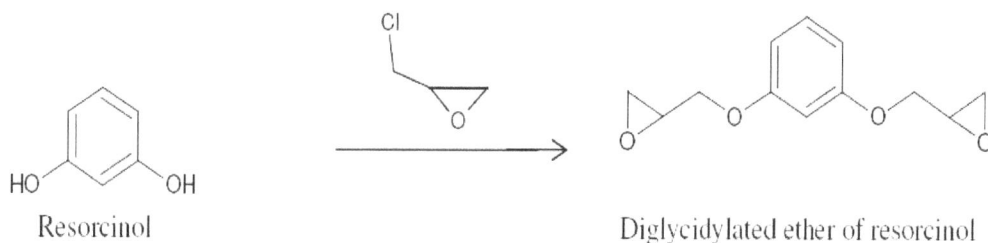

Figure 7 Glycidylation reaction of resocinol with epichlorohyd

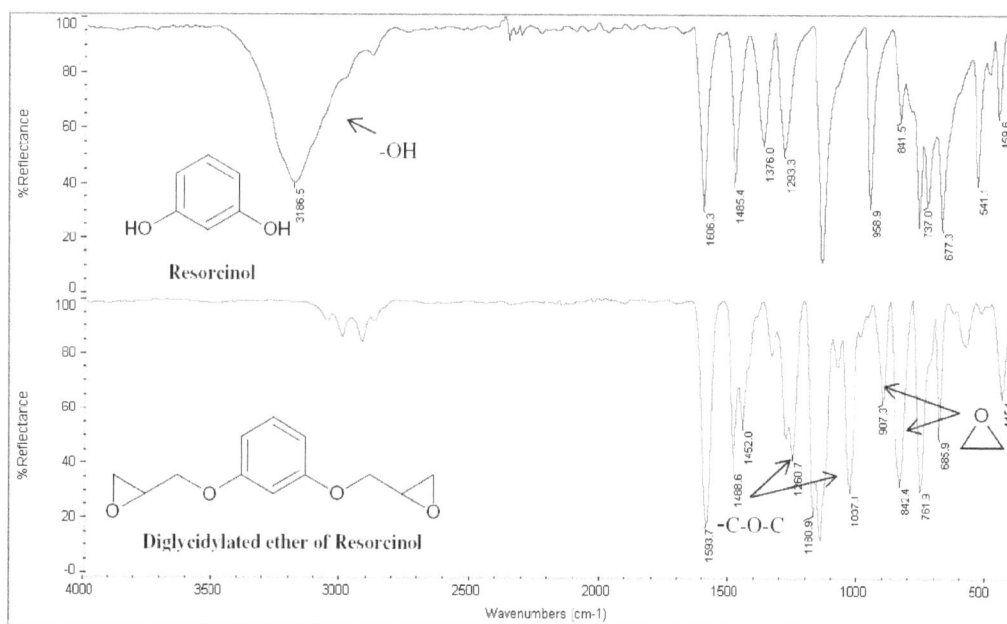

Figure 8 FTIR spectra of resorcinol and epoxidized resorcinol

THERMO-MECHANICAL CHARACTERIZATION OF BIO-BASED EPOXY THERMOSETTING RESIN

Polymerization of epoxidized linseed oil through step-growth process

The first approach to study viability of chemical modification of the vegetable oils, as well as its effect on the resulting thermos-mechanical properties, included the introduction of epoxy groups on the fatty acid chains of linseed oil. The storage modulus (E') and the glass transition (tan δ) of thermally cured epoxidized thermosets were used as a guide to evaluate the thermo mechanical performance. The epoxy/oxirane group is characterized by its reactivity toward both nucleophilic and electrophilic species, and it is thus receptive to a wide range of reagents. Epoxy monomers polymerize through step-growth and chain-growth processes. The most typical way to polymerize epoxy monomers through step growth mechanism is the reaction with amines, which are the most common curing agents to build up epoxy networks.

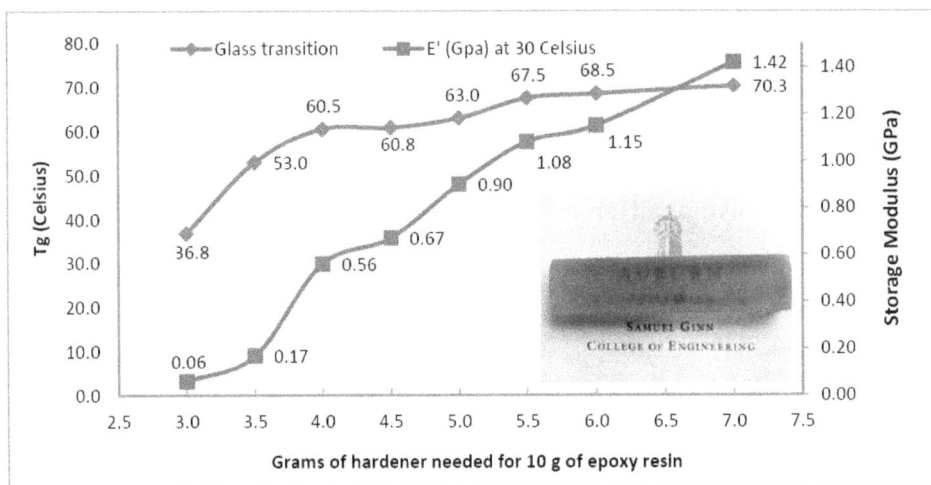

Figure 9 Storage modulus and tan δ of epoxidized linseed oil as a function epoxy/amine ratio

The epoxidized oil was mixed with a commercial hardener SC-15 part B (by Applied Poleramic Inc.), composed of cycloaliphatic amine and polyoxyl-alkylamine in different weight ratios. Previous experiments performed by this research group revealed that the best thermo-mechanical properties were obtained when using SC-15 part B as a hardener to cure the epoxidized linseed oil. Other tested commercial hardeners included Jeffamine T-403 and Jeffamine D-2000. The cure reaction proceeds following a condensation procedure. The mixture was cured for 6 hrs. at 60°C, 6 hrs. at 80°C, 2 hrs. at 120°C and finally 1 hr. at 180°C. Figure 9 shows the evolution of the storage modulus and the tan δ as a function of the epoxy/hardener ratio.

Basically, within the testing range, the epoxidized linseed oil thermosetting resin showed the maximum glass transition temperature for epoxy/hardener ratios above 10/5.5, and the highest modulus when the ratio was 10/7. An optimized Tg and modulus, trying to minimize the amount of hardener, is attained when using an epoxy/amine ratio of ~10/6. Materials with low amine proportions contain unreacted groups or even free monomers. These unreacted species are trapped inside of the polymer network and act as plasticizers lowering both the modulus and Tg.

PREPARATION OF THE LINSEED OIL BASED POLYURETHANE

The second approach comprised the modification of the linseed oil structure by introducing secondary –OH groups that can be polymerized by means of a reaction with isocyanates. The linseed oil-based polyurethane was prepared by mixing the resulting polyol with aromatic TDA.

As a first approach to evaluate the thermo-mechanical behavior of this polyurethane, a 1.0/1.1 molar ratios of the -OH group to the isocyanate group was chosen for the formulation. Using Equation a polyol equivalent weight = 244 g/eq. was obtained. Storage modulus (E`) and tan δ were recorded as a function of the temperature. Results revealed the presence of a crosslinked network, with a glass transition temperature of around – 6.5°C and a modulus with a magnitude of approx 3 MPa at ambient conditions. This is a clear indication of the elastomeric nature of the resulting materials, and its poor capacity to withstand deformation when stress is applied.

POLYMERIZATION OF ACRYLATED SOYBEAN OIL THROUGH FREE RADICAL MECHANISM

The third approach covered the study of the acrylated oils. Acrylated soybean oil was employed to explore the feasibility of using oils modified with acrylate groups to generate suitable materials for different polymer applications. The supplier states that the product has a molecular weight of approximately 1800-7200 g/mol; and a viscosity at 25°C of 18000-32000 cps. With the aim of improving the solubility of the radical initiator (AIBN), the acrylate soybean oil was diluted with methyl methacrylate, a low molecular weight comonomer. DMA was also used to evaluate the thermos-mechanical properties. Materials resulting from the polymerization of acrylated soybean oil with AIBN appeared as relatively transparent, yellowish brittle and rigid polymers.

As seen in the table, was not deeply impacted by the comonomer content and ranged from 50 to 60°C. On the other hand, storage modulus was almost duplicated when adding 25 wt% of methyl methacrylate. Higher proportions of the comonomer were not tested to minimize the presence of commercial synthetic agents. The 25 wt% methyl methacrylate proportion allowed to adequately dissolve the initiator and to reduce the viscosity, vastly improving the manipulation of the resin during formulation. The rubbery moduli of the systems ranged from approximately 30 to 45 MPa, and crosslinking densities ranging from approx. 5246 to 3461 mol/m3, when increasing the methyl methacrylate proportion from 10 to 25 wt%. The n value, as expected, decreased with the increase in the methyl methacrylate content, due to its chain extender effect (as reflected in the increase in the Mc values), by reducing the crosslinking density through the formation of linear chains embedded in the acrylated oil network. Physical densities were around 1.08 g/cm3, and did not statically changed as a function of the methyl methacrylate proportion

Table 3 Properties of acrylated soybean oil polymers

Acrylated Oil (wt%)	Methyl Methacrylate (wt%)	E` at 30 °C (GPA)	Tg (°C)	n (mol/m3)	Mc (g/mol)
90	10	0.38	50	5246	206
85	15	0.48	55	4445	243
80	20	0.67	60	4290	251
75	25	0.75	59	3461	312

As a summary of results, so far three chemical modifications (epoxy, hydroxyl and acrylate) have been preliminary evaluated. As seen, the step-growth process employed to formulate the epoxy resin implies the use of a commercial amine to crosslink the epoxy monomers. This generates a partially bio-based (around 60 wt% of the total weight of the formulation) resin that showed a considerably better thermo-mechanical performance, where non-modified tung oil was cationically polymerized to yield elastomeric materials. Clearly, this modification improves the mechanical properties of oil-based materials, yielding thermosets with modulus in the order of GPa and glass transitions temperatures around 70°C.

Concerning the second approach, the inclusion of hydroxyl groups to produce polyurethanes; it is evident that the elastomeric nature of these resins cannot make them suitable candidates to replace materials intended to structural or composite applications. Due to their flexibility, these polyurethanes could be used as matrices for thermal insulation, packing or cushioning. However, the resulting materials were extremely flexible, and lacked of mechanical resistance; therefore, no further investigations were carried out using polyurethanes. Regarding acrylated oils polymerized through free radical mechanism, it can be concluded that acrylated oils can be used to make rigid thermosetting polymers and composites. These materials exhibited high crosslinking densities, resulting in higher Tg (approx. 60°C) and improved storage modulus (~0.7 GPa) than their non-functionalized counterparts. From these findings, the epoxy and acrylate moieties are promising approaches to more reactive monomers able to generate materials with good mechanical properties than their non-modified counterparts.

It is clear however, that all of the above mentioned modifications lead to not fully, partially biobased resins; which still lacking of the desire properties to simulate the performance. Therefore, this chapter also studies the chain-growth copolymerization (which reduces the amount of commercial agents in the formulation per weight basis) of epoxy monomers. The details of the copolymerization of epoxidized linseed oil and a low molecular weight aromatic epoxy comonomer are summarized in the following sections.

Copolymerization of epoxidized linseed oil and epoxidized δ-resorcylic acid through chain-growth process

Another possible ways to polymerize epoxy monomers is through the use of small amounts of Lewis bases that act as initiators of the anionic polymerization. Commonly used initiators include tertiary amines such as 4-dimethylaminopyridine (DMAP). The mechanism for the anionic polymerization of epoxy monomers initiated by tertiary amines is highly complex and has not been fully described yet. Figure 10 depicts a simplified scheme of the anionic polymerization of epoxy monomers.

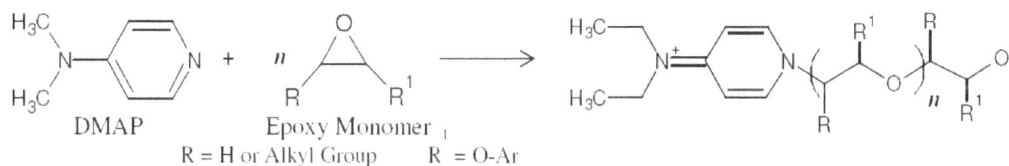

Figure 10 Anionic polymerization of epoxy monomers.

In order to prepare the resins, epoxidized linseed oil and triglycidylated ether of δ-resorcylic acid (TRA) or epoxidized δ-resorcylic acid, were mixed in different weight proportions. The idea of using this aromatic trifunctional monomer lies upon the principle that the flexibility of the polymer backbone may be markedly decreased with the insertion of groups that stiffen the chain by impeding rotation, since more thermal energy is then required to set the chain in motion. The presence of rigid aromatic structures confers to the materials the thermo-mechanical properties required in their industrial and high performance applications. Dimethylamino pyridine or DMPA was added to the mixtures in a ratio of 0.08 mol DMAP/mol epoxy. After the curing process, the polymeric materials appeared as dark-brown stiff materials at room temperature. For all the studied systems, suitable materials for analysis were not obtained for epoxidized linseed oil proportions above 30 wt%; due to the lack of miscibility of the two compounds.

CONCLUSIONS

This chapter focuses on the chemical modification of two different functional groups of biomass derived materials; the carbon-carbon double bonds of triglycerides, and the hydroxyl groups of phenolic compounds. These chemical modifications were addressed to introduce more reactive functional groups able to polymerize and to produce materials with a better thermo-mechanical performance. Three chemical modifications (epoxy, hydroxyl and acrylate) were preliminary evaluated. The step-growth process employed to formulate the epoxy resin implies the use of a commercial amine to crosslink the epoxy monomers. It was observed that the epoxidized oil, when mixed in an epoxy/amine ratio of ~10/6, produced resins with acceptable thermomechanical properties, showing a Tg of around 70°C and a modulus of around 1.4 GPa at room conditions. Regarding acrylated oils polymerized through free radical mechanism, it can be concluded that acrylated oils can be used to make rigid thermosetting polymers and composites. These materials exhibited high crosslinking densities, resulting in higher Tg (approx. 60°C) and improved storage modulus (~0.7 GPa) than their non-functionalized counterparts. These two approaches produced materials with considerably better thermo-mechanical performance than the thermosets.

PREPARATION AND CHARACTERIZATION OF A SOY PROTEIN BASED BIO-ADHESIVE

INTRODUCTION

Industrial interest in bio-based polymeric materials from renewable agricultural products has received increased attention over the past decade. Soy proteins are among the most investigated natural compounds for various industrial applications such as wood adhesives, food packaging, bio-based composites and coatings, etc. due to their plentiful supply, excellent properties and high functionality. In recent years, soy proteins have attracted great attention to prepare bio-adhesives for bonding wood composite materials.

However, the low bonding strength and poor water resistance of soy protein bioadhesives limited their more extensive application in highperformance wood adhesives. Soy protein, the main byproduct obtained from the production process of soybean oil, contains several functional groups, such as primary amine, hydroxyl and carboxylic groups. These reactive function groups make the soy protein molecule available to be modified by chemical approaches for preparing a competitive formaldehyde-free bioadhesive.

Thus, many efforts have focused on improvement in the adhesive properties by modifying the soy protein, mainly including chemical denaturation, crosslink modification, enzyme treatment, and protein molecular modification.

The denaturation treatment of soy protein usually used urea, alkali, sodium dodecyl sulfate, and guanidine hydrochloride to unfold the complex molecule structure of protein and expose the internal hydrophobic groups to enhance the soy protein adhesives water resistance. However, the resulting molecule structure network of adhesive is relatively easy to hydrolyze in wet conditions, due to its contained many hydrophilic groups.

For soy protein molecular modification, high activity groups were normally grafted onto soy protein molecules through esterification, acylation reaction, etc. to form a cross-linked network structure. This method can markedly improve the water resistance of soy protein adhesive; however, the main drawback is the complex preparation procedures, which makes it impractical for application in the plywood production.

Crosslinking modification is regarded as an effective and facile approach for improving the water resistance of soy protein adhesives. This modification means use of different chemical species to increase the spatial density of potential crosslinking site. Active polymer and resin appear to be efficient and feasible cross-linkers, since their containing reactive groups can react with the hydroxyl, amino and carboxylic acid groups of soy protein molecular to form a covalent-linkages network within the protein matrix to improve the bonding properties and water resistance of soy protein adhesives. Several researchers have used polymeric methane diphenyl diisocyanate, polyamidoamine-epichlorohydrin (PAE) resin, polyethylene glycol adipate, synthetic latex and epoxide resins as cross-linkers to modify soy protein adhesives, and they have already been proven effective. In these cross-linkers, epoxy monomer and resulting resins have received special attention, because epoxy groups can freely react with abundant nucleophilic groups such as $-NH2$, $-OH$ and $-COOH$ on the soy protein molecular.

Furthermore, most epoxy resins were produced from the epichlorohydrin, which can be derived from sustainable resource of bio-based glycerol. Huang J. et al. synthesized three types of polyepoxides including triglycidylamine (TGA), glycerol polyglycidyl ether (GPE) and trimethylolpropane triglycidyl ether (TTE) as crosslinker for soy-based adhesives. The plywood panels bonded by soy-based adhesives with TGA and GPE did meet the interior use requirement.

Commercial epoxy resin was also an effective cross-linker for enhancing the water resistant and bonding properties of soy protein-based adhesives. Luo et al. reported that a soy-based adhesive was prepared by using soybean flour, polyacrylamide, and epoxy resin as cross-linker, which met wet shear strength requirements of interior use panel.

Enzyme treated soy-based adhesive was blended with commercial epoxy resin before the application, and a water resistance cross linked structure was formed during its curing process, resulting in the improvement of plywood bonding strength. In addition, the addition of epoxy resins to soy-based adhesives was found to improve its gluability and flowability. These efforts mainly focus on physical mixing of soy-based adhesive directly with cross-linker before its application, whose water-resistant structure was formed during its hot-press curing process.

However, taking account of longer curing time of epoxy resin, the limited hot-pressing time was not enough for the sufficient cross linking reaction between epoxy groups and functional groups of soy protein. Furthermore, epoxy resin is oil soluble, which made it difficult to adequately dissolve in water-based soy protein adhesives, thus resulting in the reduction of reinforcement efficiencies of epoxy resin and requiring a large dosage.

Waterborne epoxy is dispersed in aqueous solution as colloidal particles, and amine is usually utilized as curing agent. More importantly, many waterborne epoxies are formulated without volatile solvents and use merely water as a diluent, which leading to ultra-low or even zero, VOC paints and varnishes. Therefore, using waterborne-epoxy resin as cross linking agent added during the preparation process of soy protein adhesive through effective cross linking provide a simple and useful approach to form uniform and well distributed bio-adhesive. To our knowledge, this approach to soy protein adhesives modification has not, to date, been reported.

In the present work, a series of soy protein bio-adhesive was developed using waterborne-epoxy resin as cross-linker and soy protein. Because the reactive functional groups on soybean protein molecule are limited, we introduced polyacrylamide (PAM) containing rich amino group into the adhesive matrix to increase cross linking density through the multiple reaction of epoxy resin, soybean protein, and PAM. Effects of epoxy resin and PAM on the developed adhesive properties were investigated, and by measuring initial viscosity and wet shear strength of plywood verify the performance of soy protein adhesives were enhanced. Cross linking reaction between waterborne-epoxy resin and soy protein was also confirmed by 1H nuclear magnetic resonance (NMR). Differential scanning calorimetry (DSC) was also employed to evaluate curing properties of the adhesives.

MATERIALS AND METHODS

Soybean protein isolate (SPI, $95% protein) was purchased from Baiwei Biological Technology Co., Ltd, Hebei, China. Defatted soy flour (SF) with about 52.2% of crude protein ground from defatted soy meal was purchased from Shandong San Wei Oil Group Co., Ltd. The remaining components in SPI are mainly moisture (#4%), ash (#3%), and fiber (#1%). The SPI was ground to 200 mesh or more firstly. Sodium hydroxide (NaOH) and other chemical reagents were provided by Xilong Scientific Co., Ltd. Non-ionic polyacrylamide (PAM) with 1.2 _ 107 molecular weights was obtained from Ruida Purification Materials Co., Ltd., Henan, China. A commercial waterborneepoxy resin of bisphenol-A (F0704, without curing agent, the storage life is 6 months, and should be stored in a dry and clean house) was purchased from Shenzhen Jitian CHEMICAL Co., Ltd. from Guangdong Province, which had a weight per epoxide (WPE) of 463–468, 50% of solid content and viscosity of 820–950 mPa s. Poplar veneer (400 mm - 400 mm - 1.7 mm, 10% moisture content) was supplied by Linyi, Shandong Province, China.

PREPARATION OF SOY PROTEIN ADHESIVE

Soy protein adhesive crosslinked by epoxy resin (EPOXY-SPI) and PAM modified EPOXY-SPI (EPOXY-SPI-PAM) were prepared according to the following procedure:

First, for the pure SPI adhesive, 20 g of SPI and 170 g of distilled water were added to into a three-necked flask and stirred rapidly for 20 min at 40°C water bath. And then the pH value of SPI slurry was adjusted to about 9.0 with 20% NaOH solution (w/w).

For the EPOXY-SPI adhesive, various mass fractions of epoxy resin (1%, 2.5%, 4%, 5.5% or 7% of the pure SPI adhesive mass, based on the solid content of epoxy resin) were added into the pure SPI adhesive and continued stirring 30 min at 45°C. Final adhesive was then obtained after cooling reaction product temperature to 25–30°C.

For the EPOXY-SPI-PAM adhesive, the same procedure and the same amounts of ingredients as EPOXY-SPI adhesive were used except that different concentrations of PAM aqueous solution were used to substitute for distilled water, and the addition amount of epoxy resin was set to 4%. The used amounts of PAM (0.05, 0.1, 0.3, 0.5 and 0.7 wt%) were determined by the relative weight percentage of the resulted adhesive.

As shown in Fig. 1, the spectra of SPI and EPOXY-SPI are obviously different

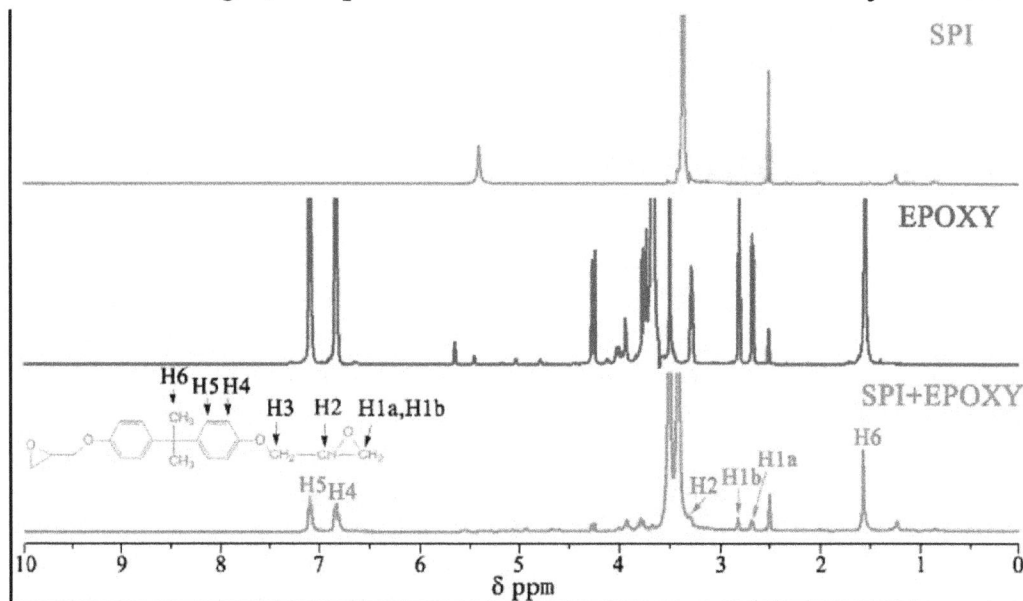

Fig. 1 The 1H-NMR spectra of different adhesives

The appearance of new chemical shifts in 1H-NMR spectra is powerful evidence regarding whether or not cross linking reaction between epoxy resin and SPI takes place during the preparation process of EPOXY-SPI adhesive. Therefore, 1H-NMR analysis was performed, and the spectra of the SPI, EPOXY resin and EPOXY-SPI adhesive are shown in Fig. 1, which is in consistent with reported analytical data for similar structures.

The likely reaction mechanism was presented in Fig. 2.

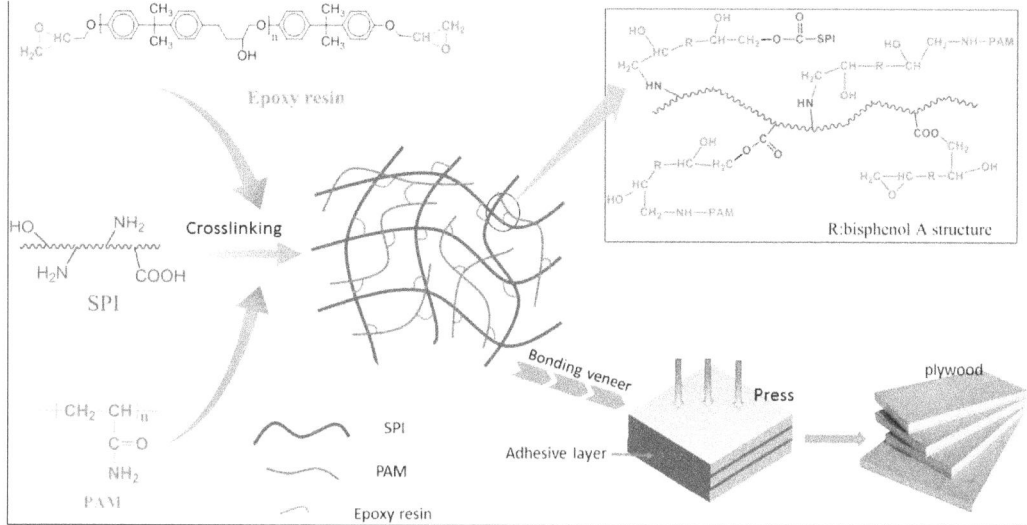

Fig. 2 The reaction mechanism among of SPI, epoxy resin and PAM

CONCLUSION

In this work, an environmentally friendly soy protein bioadhesive was developed using waterborne-epoxy resin, soy protein, and water-soluble PAM. Epoxy resin was crosslinked with the active groups of soy protein as well as PAM to form a water-resistance network structure through their multiple reactions. 1H-NMR analysis results also indicated that the resulting adhesive contain a certain amount of epoxy groups, which can further react with hydroxyl group of wood to form better bonding property during its curing. It is also seen that the addition of PAM to EPOXY-SPI was capable of decreasing the cure temperature and increasing the wet shear strength of soy protein adhesives. The crosslinked adhesive with the dosage of 4 wt% epoxy resin and 0.05 wt% PAM showed better wet shear strength and water resistance property. When the epoxy resin addition increase from 5 wt% to 10 wt%, the property of EPOXYSF- PAM adhesive is also meet requirement of interior use plywood. In view of the stringent formaldehyde emission limits imposed on interior-used plywood, the waterborne-epoxy resin crosslinked soy protein adhesive is a better substitute as wood adhesive for formaldehyde-based resins.

18
Chapter

POLY (LACTIC ACID) FILLED WITH CASSAVA STARCH-G-SOYBEAN OIL MALEATE

INTRODUCTION

The petroleum-derived plastics have been used extensively, causing tons of thousands of plastic waste. The improper disposal of petroleum-derived plastics leads to environmental pollution which has aroused much interest in searching for substitutes. Recently, biodegradable and renewable polymers have been increasingly developed. Generally, polymers from renewable resources can be classified into three groups: (1) natural polymers such as starch, protein, and cellulose; (2) synthetic polymers from natural monomers such as polybutylene succinate (PBS) and polylactide (PLA); and (3) polymers from microbial fermentation such as polyhydroxybutyrate.

These polymers are aliphatic polyesters which are biodegradable and compostable thermoplastics derived from renewable resources, such as starch and sugar cane. PLA and PBS are the most promising thermoplastic polymers in this regard. They are compostable which is perfectly suitable for consumer goods and packaging applications. They are derived from renewable resources such as corn, sugar cane and cassava. They are biodegradable polymers which are nontoxical and acceptable mechanical performance but their applications, particularly PLA, are limited by the high cost and relatively poorer propertieswhen compared to petroleum-based plastics. PBS is semicrystalline polymer which exhibits the flexibility when compared to PLA. On the other hand, PLA is brittle polymer due to its high Tg, low crystallinity and low thermal stability.

With respect to its chemistry, PLA is synthesized by direct condensation polymerization of the lactic acid monomers or by ring opening polymerization of lactide monomer. There have been several possibilities to modify its property such as the nucleating agent addition, polymer composites, and polymer blend preparation with natural fillers such as starch or cellulose and with natural rubber, respectively. The addition of starch into PLA is one of the most promising efforts due its abundant and cheap biopolymer particularly cassava starch since it is the fifth most abundant starch crop produced in the world and the third most important food source for inhabitants of tropical regions.

As a result, starch has been used as filler for environmentally friendly plastics. However, the challenge is the major problem associated with poor interfacial adhesion between the hydrophilic starch granules and the hydrophobic PLA, leading to poorer mechanical properties. Therefore, several strategies have been adopted to improve the compatibility such as by using compatibilizers or reactive coupling agents.

Especially, maleic anhydride- (MA) based compatibilizing agents are the most popular due to their good chemical reactivity, low toxicity, and low potential to polymerize itself under free radical grafting conditions. Wang et al. prepared thermoplastic dry starch (DTPS)/PLA blends by using MA as compatibilizer and dicumyl peroxide (DCP) as initiator to enhance the compatibility between DTPS and PLA. Glycerol was employed as a plasticizer for dry starch to avoid the depolymerization of hydrophobic PLA during melt processing. The plasticization of starch and its compatibility modification with PLA was accomplished in a single-screw extruder by one-step reactive extrusion. The results have been reported that the physical properties and compatibility of DTPS/PLA blends were improved.

This research was focused on the effect of soybean oil maleate grafted cassava starch as filler on the properties improvement of PLA. The soybean oilmaleate grafted cassava starch was easily synthesized by the maleation of soybean oil with maleic anhydride. Thus, obtained soybean oil maleate grafted cassava starch was mixed with PLA by twin screw extruder using various weight ratios.

MATERIALS AND METHODS

PLA pellet (2003D grade) by Nature Works was purchased from BC Polymers Marketing Co., Ltd. (Bangkok, Thailand). It is transparent polymer with a glass transition temperature (Tg) of 58–60°C and a density of 1.24 g/cm3, as reported by the manufacturer. Cassava starch flour was purchased from Thai Wah Food Products Public Company Limited (Bangkok, Thailand). Soybean oilwas supplied from Siam Chemical Industry (Thailand).Maleic anhydride (≥98% purity) produced by Sigma Aldrich, dicumyl peroxide (98% purity) manufactured by Sigma Aldrich, and acetone (AR grade) were purchased from the local distributor (FACOBIS Co., Ltd.).

Preparation of Soybean Oil Maleate: Soybean oil (100 g) was charged into a 500 mL glass reactor. Then, 10g or 20g of MA powder (two reaction batches; batch one: MA 10wt% based on soybean oil and batch two: MA 20wt% based on soybean oil) was added and continuously stirred. Following that, DCP (3 wt% of MA) was used as a free radical initiator. The reaction was allowed to continue at temperature of 170°C for 2 hrs. The viscous yellowish liquid of soybean oil maleate (SOMA) was obtained. SOMA10 and SOMA20 were assigned to soybean oil maleate having anhydride content of 10wt% and 20wt%, respectively.

Preparation of Cassava Starch-g-Soybean Oil Maleate: 50g SOMA and 50g starch powder at weight ratio of 1:1 were well mixed using a mechanical stirrer. After that, the mixture was put in the hot-air oven and the temperature was set at 120°C for 2 hours. It was anticipated that at this condition, SOMA underwent the ring opening reaction with starch hydroxyl group, yielding SOMA-g-STARCH. Then, the crude product was washed with acetone to remove unreacted SOMA. Two sorts of SOMA-g-STARCH were prepared which were assigned to SOMA10-g-STARCH and SOMA20-g-STARCHfor SOMA-g-STARCHs prepared from SOMA10 and SOMA20, respectively. The obtained SOMA-g- STARCHs were dried in an oven at 50°C and ground prior to subsequent experiment.

Preparation of PLA Filled with Cassava Starch-g-Soybean Oil Maleate: The obtained SOMA-g-STARCH (SOMA10-g- STRACH and SOMA20-g-STRACH) was mixed with PLA in various PLA: SOMA-g-STARCH weight ratios of 90 : 10, 80 : 20, 70 : 30, 60 : 40, and 50 : 50 using twin-screw extruder and the barrel zone temperature was set at 135, 160, 170, 180, and 180°C fore zone 1, zone 2, zone 3, and zone 4, respectively.

Characterization and Testing: Fourier transforminfrared spectroscopy was conducted using Nicolet FTIR spectrometer (Nicolet 6700) recorded in the range of 3500–1000 cm-1. Proton NMR was recorded on Varian UNITYINOVA 500Hz spectrometer using deuterated acetone as a medium. Morphology was observed by a JEOL, JSM-6480LV scanning electron microscope (Tokyo, Japan) using acceleration voltage of 22 kV. Colored Appearance was evaluated by dyeing 1wt% red disperse dye using H24 Newave Lab Equipment (Taiwan). TGA analysis was conducted on Mettler Toledo TGA/SDTA851e thermalgravimetric analyzer (Columbus, Ohio) which was carried out under nitrogen atmosphere with the flow rate of 20 mL/min and heating from 25°C to 600°C at heating rate of 10°C/min. Impact strength was determined by izod impact tester (GOTECH) according to ASTM D256 standard, and tensile properties were determined by universal testing machine (LR 100 K) according to ASTM D638 standard and using load cell of 100 kN and crosshead speed at 50 mm/min.

RESULTS AND DISCUSSION

Characterization of Soybean Oil Maleate: The graft reaction between soybean oil and maleic anhydride initiated by DCP to produce SOMAis proposed by Scheme 1.The appearance of SOMA is viscous and brownish. Upon standing for several days, the unreactedMA is crystallized and discarded, leaving only SOMA liquid. The SOMA functional groups were identified by FTIR analysis of which FTIR spectra.

The typical absorption band of carbonyl ester group of soybean oil is found to be 1720 cm-1. From the spectra of SOMA10 and SOMA20, the additional absorption bands at 1775 cm-1 and 1850 cm-1 are observed, corresponding to the symmetric and asymmetric stretching of C=O in the pendent anhydride group.

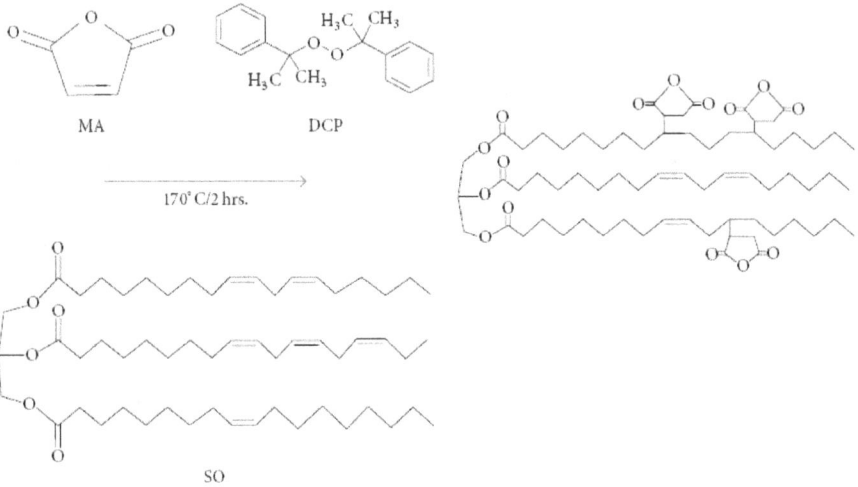

Scheme 1: Representative grafting reaction of soybean oil with maleic anhydride using dicumyl peroxide as an initiator.

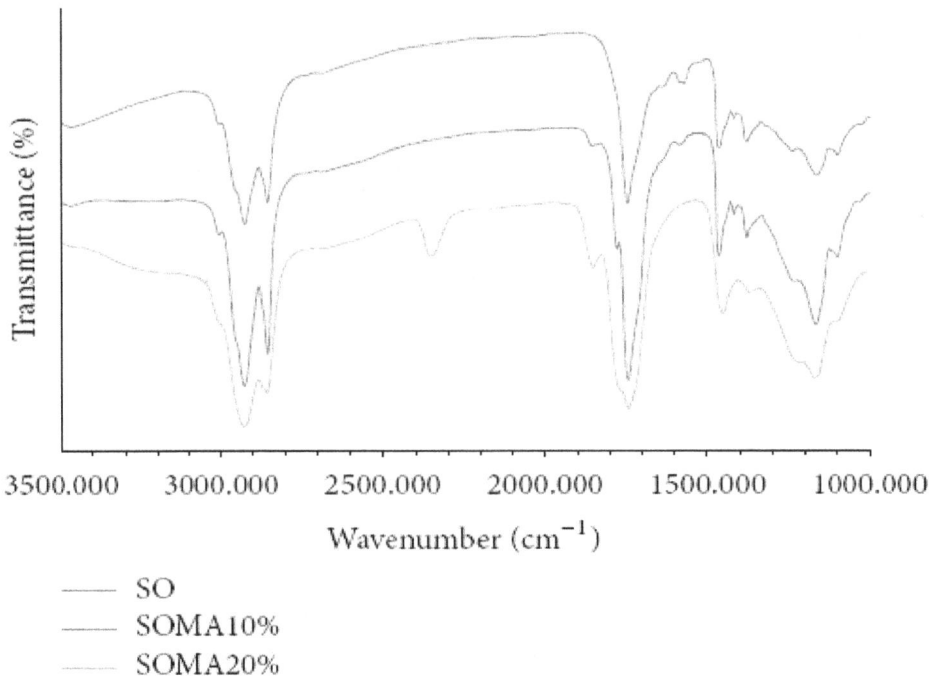

Figure 1: FTIR spectra of soybean oil and soybean oil maleate.

Their absorption intensities tend to increase with an increase in the amount of applied MA, indicating that the more the amount of added MA the more the amount of anhydride ring content. This anhydride ring remains intact since no absorption peaks related to –OH stretching in the region of 3,500 cm-1 due to the opening of the anhydride were observed. The presence of anhydride ring is important for the subsequent grafting reaction.

The 1H-NMR technique was employed to confirm the structure of SOMA. The NMR spectra of SOMA and SO. The signals of methylene protons of anhydride pendant are obviously seen at around 2.60–2.70 ppm(a, b) and at around 2.80–2.90 ppm (c) which are not present in the spectrum of SO. Based on two techniques, it is certain that SOMA was successfully synthesized and the possible structure is proposed as shown in Scheme 1.

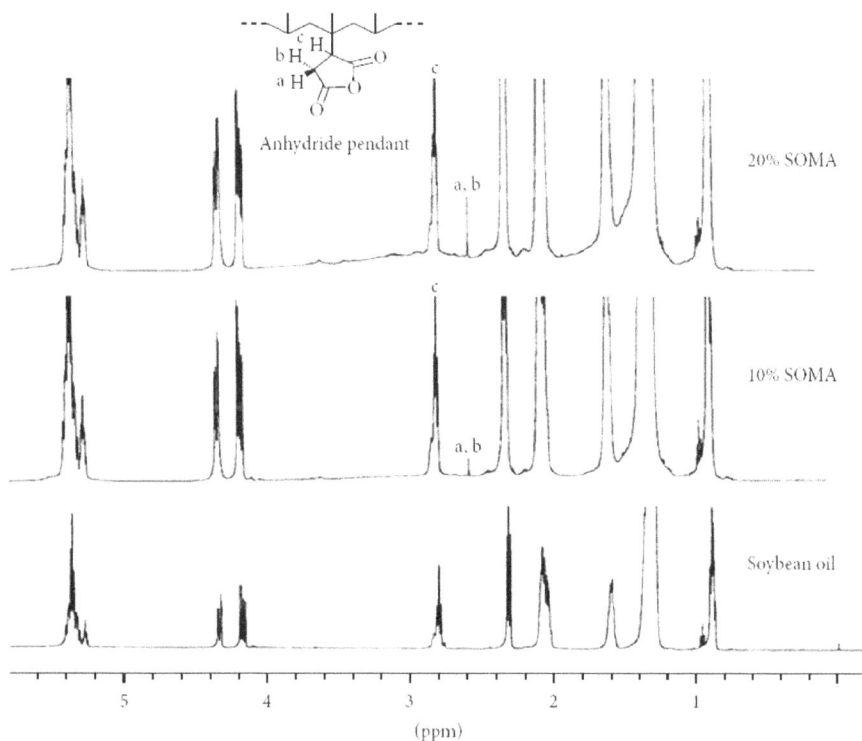

Figure 2: 1HNMR spectra of soybean oil and soybean oilmaleate.

CONCLUSIONS

Maleated soybean oil whose functional group and chemical structure were confirmed by FTIR and 1H NMR was successfully prepared by grafting reaction of soybean oil with maleic anhydride using dicumyl peroxide as an initiator. Then, maleated soybean oil was employed for surface modification of cassava starch powder, producing soybean oil grafted starch powder (SOMA-g-STARCH). FTIR analysis and TGA analysis confirmed that SOMA-g-STARCH surface changed from hydrophilicity to hydrophobicity.

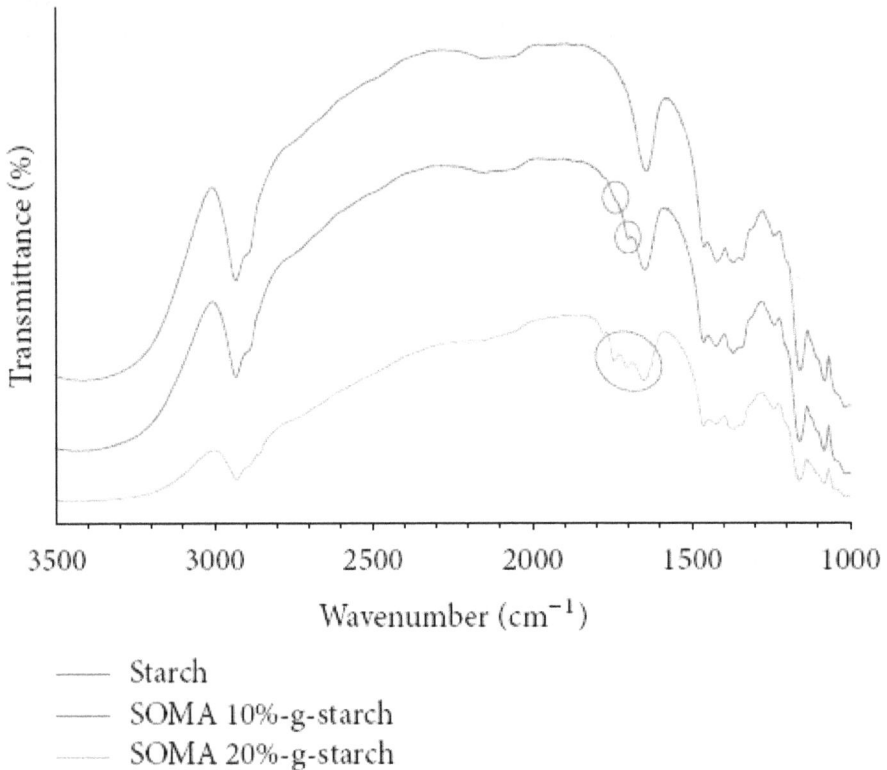

Figure 3: FTIR spectra of starch and soybean oil-g-cassava starch powder.

The incorporation of SOMA-g-STARCH into polylactic acid by melt extrusion mixing using PLA: SOMA-g-STARCH weight ratios of 90 : 10, 80 : 20, 70 : 30, 60 : 40, and 50 : 50 was carried out. It was found that the loading of SOMA-g-STARCH into PLA resulted in a significant improvement of impact strength, in case of SOMA-g-STARCH contents which have below 30 wt% due to better interfacial adhesion and good particles distribution. However, PLA/SOMA-g-STARCH composites with SOMA-g-STARCH content higher than 30 wt% exhibited a decrease in the impact strength, arising from the problem of particle agglomeration. In a similar manner, polymer composites of PLA filled with cassava starch-gsoybean oil maleate exhibited higher mechanical properties than polymer composites of PLA filled with unmodified starch powder. It was thought that soybean oil on the particle surface might play a key role in improving the compatibility as well as performing the plasticity-like behavior.

19
Chapter

POLY (L- LACTIC ACID) FILLED WITH CASSAVA STARCH-G-SOYBEAN OIL MALEATE

INTRODUCTION

Application of natural fiber reinforced composites increases in all areas of production, especially the building and automotive industry. They are usually based on commodity polymers, like polypropylene or polyvinyl chloride. However, with an increasing environmental awareness there is a need to seek and replace traditional composites, which are difficult to recycle, with bio-based polymers and their composites, which can be easily disposed at their end of lifetime.

Several bio-based and/or biodegradable polymers have been developed commercially, such as polyhydroxyalcanoates, poly(L-lactic acid) (PLLA), polycaprolactone and polybutylene succinate and their derivates . Among those listed, PLLA gained the greatest attention due to its numerous advantages such as being biodegradable and compostable, having good stiffness and strength and being able to be produced on a large scale by microbial fermentation of agricultural byproducts . Besides its advantages, PLLA has also some drawbacks, mainly brittleness and slow crystallization kinetics that should be overcome to further develop its application possibilities. Several attempts have been made to outbalance these disadvantages, which mainly include the modification of PLLA formulations to develop material with improved plasticization, impact resistance, or crystallization rate, often achieved by introduction of particles and fiber-like fillers.

The incorporation of natural fibers into PLLA results in price reduction, as natural fibers are typically industrial biowaste. Previous literature studies indicate that the addition of natural fibers to PLLA leads to compostable materials with increased stiffness and even an accelerated biodegradation process due to faster hydrolysis and oxidation of the polymer matrix and filler in specific environment. During the last two decades, researchers extensively studied PLLA-based composites attained with a vast variety of natural fibers, such as: wood, cotton, nut shells, bamboo, cereal straw, jute, sisal, and abaca.

Recently, attention has been focused on PLLA/wood composites, due to their promising properties and potential applications. The type of wood, namely its specific chemical composition, may affect the possible interaction with the matrix. Pine wood, which is the filler selected for our study, is mainly composed of cellulose (40–50 wt %), lignin (25–30 wt %), hemicellulose (20–25 wt %), and small amounts of fatty acids and resins (up to 10 wt %).

The building blocks of pine wood are microfibrils made of highly crystalline Cellulose acting as the frame, surrounded by semicrystalline hemicellulose as the matrix and covered by amorphous lignin. These microfibrils have a diameter of 10–20 nm and make up the cell walls of the tree.

Previous literature studies indicate the need to improve interfacial adhesion between the wood filler and PLLA. Wood particles have often a large size, of the order of hundreds of microns, which easily deboned from the polymeric matrix under external load. Deboning causes formation of voids and in turn premature failure of the material, as it was demonstrated for instance in the deformation process of polypropylene/wood composites not containing any coupling agent. A number of coupling agents were tested to improve interfacial adhesion between PLLA and wood, with only limited results in terms of improved compatibility, as for N,N-(1,3-phenylene dimaleimide) and 1,1-(methylenedi-4,1-phenylene) bismaleimide, phenolic resin, or bioadimide.

Silane coupling agents were also tested, with effciency found to be affected by the specific functional groups. α-methacryloxypropyltrimethoxysilane was used as a compatibilizer in PLLA/pine woodflour composites, with poor results in terms of even increased brittleness compared to PLLA.

An improvement of the thermal and mechanical properties of PLLA reinforced with Stika fibers treated with vinyl trimethoxysilane was reported. A direct comparison of the efficiency of silane functional groups was conducted for unsaturated polyester and epoxy resin matrices filled with silane-treated cellulose fibers, using α-aminopropyltriethoxysilane (APE), α-methacrylopropyltrimethoxysilane, hexadecytrimethoxysilane, and α-mercaptopropyltrimethoxysilane as coupling agents. An increase of the modulus and tensile strength in the resulted composites was observed, with the strongest reinforcing effect found for APE-treated fibers, and rationalized with higher reactivity of the (–NH2) functions. APE was also found to be an efficient coating agent in PLLA/cellulose fibers and nanofibers composites, by promoting dispersion of cellulose nanofibers, with improved adhesion between the phases and increased storage modulus in PLLA/cellulose fibers composites.

APE could also promote compatibilization between PLLA and lignin, the other major constituent of pine wood, when used as coating for lignin. These literature results suggest high potential for APE to act as an efficient coupling agent also for PLLA composites containing pine wood, where, as mentioned above, cellulose and lignin are the major constituents. To our knowledge, to date APE has not been tested yet as a compatibilizer for PLLA/pine wood composites. Numerous studies have been conducted on PLLA/wood blends in order to investigate their molecular structure, crystallization kinetics, thermal, and mechanical properties Researchers also focused on testing different PLLA formulations using solid-state NMR techniques, mainly high-resolution 13C and 1H NMR spectra and carbon spin-lattice relaxation times T1 (13C).

However, according to the best of our knowledge, very limited research describes molecular dynamics of PLLA-based modifications using the temperature studies of spin-lattice relaxation times (T1). This type of the solid-state NMR technique describes molecular dynamics of local and segmental motions of side groups and polymer chain segments and allows for determination of diffusion coefficients, which provide information on the size of domains in heterogeneous systems. The results presented herein provide important insights into relaxation processes of entire PLLA chains. Motions of polymer chain segments and side groups may be significantly influenced by the presence of a filler and/or compatibilizer, as they may interact via functional groups of PLLA, such as carbonyl (–CO), methane (–CH) and methyl (–CH3), with numerous side groups of lignin and cellulose, i.e., methoxyl groups (–OCH3), phenolic and aliphatic hydroxyl groups (–OH), and amino functional (–NH2) of the APE binder.

The aim of this work is to study the influence of pine wood surface-modified with APE, on structure and mechanical properties of PLLA/wood composites, with the final objective of producing completely biodegradable and bio-based wood-polymer composites.

EXPERIMENTAL

Materials

A commercial poly (lactic acid) 3100HP, abbreviated PLLA, with MFR 24 g/10 min (210OC, 2.16 kg) produced by Nature Works (Blair, NE, USA) was used. Pine wood particles were kindly obtained from Wood Technology Institute in Poznan, Poland. The as-received particles were initially selected using a Fritsch Analysette 3 vibratory sieve shaker to have a uniform diameter distribution, and then their size was measured with an Opta-Tech optical microscope (Warsaw, Poland), model MB200 combined with an Opta-Tech MI6 camera (Warsaw, Poland) and Opta-Tech Capture 2.0 software. The used particles have a length (L) of 1.25 - 0.63 mm and diameter (D) of 0.47 - 0.15 mm, and an average aspect ratio (L/D) of 2.6. Composition of pine wood particles was determined by thermogravimetry, using the procedure detailed in Ref.: the pine wood particles consisted of 62 wt % of hemicellulose and cellulose, 34 wt % of lignin and the remaining part (4 wt %) were fatty acids and resins. The chemical coupling agent was 3-aminopropylthrietoxysilane (APE) produced by UniSil (Poland), which aimed to promote adhesion between the polymer and filler.

WOOD IMPREGNATION WITH COMPATIBILIZER

APE was dissolved in ethanol (3% v/v), then the proper amount of wood to attain impregnation with 3 wt % of APE was added to the system. The filler was transferred slowly into the APE solution and stirred for 1 h at 2000 rpm. The whole impregnation procedure was conducted at room temperature. The wood filler was then filtered and dried under vacuum at 80OC for 24 h.

Composites Preparation

The composites were prepared following a four-step procedure. At first, the wood filler was modified by impregnation with APE, then the solid-state mixed with PLLA pellets, then melt mixed with the polymer matrix and finally shaped to obtain plates by compression molding. All materials were dried prior to processing at 80OC for 3 h. Poly (lactide acid) pellets were mixed with the pine wood filler in a rotary mixer Retsch GM 200 (Haan, Germany) for 3 min at a rotation speed of 2000 rpm.

Homogenization of the premixed materials with different wood contents (0–30 wt %) and compatibilizer concentration of either 0 or 3 wt % to the wood content was ensured by molten state extrusion with a Zamak corotating twin screw extruder operated at 180OC and 100 rpm. The extruded rod was cooled in air and pelletized. PLLA/wood and PLLA/wood/APE composites at various compositions were prepared, as summarized in Table 1.

Table 1. Symbols and mass concentrations of samples

Designation	Mass Concentration (wt %)		
	PLLA	Wood	Compatibilized Wood
PLLA	100	0	0
PLLA/10 W	90	10	0
PLLA/20 W	80	20	0
PLLA/30 W	70	30	0
PLLA/10 W/APE	90	0	10
PLLA/20 W/APE	80	0	20
PLLA/30 W/APE	70	0	30

The composites were compression-molded with a Remi-Plast Laboratory Forming Press PLLHS-7 (Czerwonak, Poland) at a temperature of 185OC for 3 min, without any pressure applied, to allow complete melting. After this period, a load of about 175 bars was applied for 3 min, and then the samples with a thickness of 1 mm were cooled in air to room temperature.

Results and Discussion

Heat flow rate plots of PLLA and PLLA/wood compression-molded composites are presented in Figure 1. Figure 1 illustrates the heat flow rate plots of compression molded sheets of the composites without (left) and with APE compatibilizer (right). The DSC plot of plain PLLA displays a glass transition (Tg) at 67OC coupled with sizable enthalpy relaxation, followed by a sharp cold crystallization exotherm peaked at 98OC, and by a multiple thermal event, typical of PLLA, due to transformation from -α-crystal to β-form

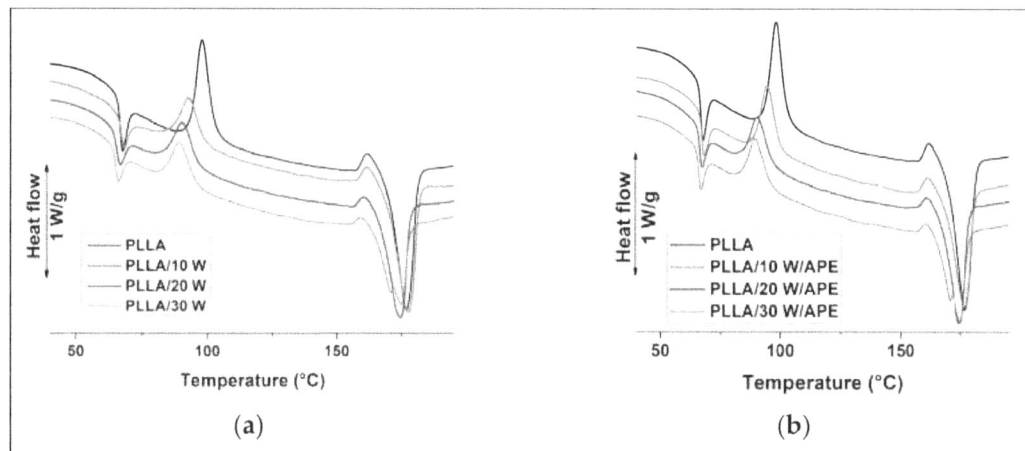

Figure 1. Heat flow rate plots of PLLA and PLLA/wood composites of the indicated compositions upon heating at 10 K/min. (a): PLLA/wood composites and (b): PLLA/wood/APE composites. Exotherm upward

CONCLUSIONS

Composites prepared with PLLA and pine wood, both bio-based, have been produced by twin-screw extrusion in different compositions, from 10 to 30 wt % of filler content, and characterized for their morphology, physicochemical and mechanical properties. Impregnation of pine wood with α-aminopropyltriethoxysilane can improve adhesion between the phases, with positive effects on material properties. The improved morphology was proven by SEM. Analysis of relaxation times by NMR indicated higher compatibility between the filler and matrix in case of surface-modified formulations. Molecular dynamics analysis proves restriction in mobility of polymer chains and CH3 side groups in case of APE-modified samples. In the non-compatibilized composites, the activation energies of CH3 groups are lower. Thermogravimetric analysis revealed that wood addition causes a decrease in thermal stability of the material, which occurred to a lower extent in the compatibilized formulations. Studies on mechanical properties indicated that the uncoated filler addition results in higher stiffness of the composites, wherein wood compatibilization leads to an enhanced connection between the matrix and filler and results in greater deformability of the entire system. Worth noting is the fact that although enhanced connection between the PLLA/wood/APE components is evident, the silanization process still needs to be improved in order to obtain satisfactory mechanical behavior that can be tolerated by practice.

20
Chapter

NEW BIO-COMPOSITES CONTAINING INDUSTRIAL LIGNINS

INTRODUCTION

Lignins are renewable and natural polymers the structure of which varies depending on plant origin and process of isolation. They are the most abundant aromatic polymers on earth and second only after cellulose. Lignin content varies from 20% to 35% depending on the plant origin. Lignins are the main components of residual liquors from chemical pulping, notably black liquor from the Kraft process. It is estimated that in Quebec, 130,000 tons of black liquor containing lignin could be used for lignin precipitation annually, without disturbing the mill production. In this investigation, two lignins were precipitated from black liquor obtained from two Quebec pulp and paper mills (Domtar Windsor and Kruger Wayagamack), following a procedure inspired by Ligno Boost protocol, using carbon dioxide to decrease the pH. A comparison of these Kraft lignins with a commercially available Kraft lignin (IndulinAT) was performed. Kraft lignins were modified by esterification in order to improve their compatibility with a polymer matrix (HDPE) and consequently the tensile properties of the bio-composites produced from that materials. In the second application, the use of non-modified and esterified lignins as coupling agent between HDPE and yellow birch (Betula alleghaniensis) or black spruce (Picea mariana) bark fibers was examined.

MATERIALS AND METHODS

Two Kraft black liquors from Quebec pulp and paper industry (Kruger Wayagamack and Domtar Windsor) were used to precipitate Kraft lignin. A commercial Kraft lignin (Indulin AT) was received from Westvaco Corporation. The high density polyethylene (HDPE) from Exxon Chemicals used in this study has a melt index of 5 g/10 min and a density of 948 kg/m3. Maleic anhydride grafted to polyethylene (MAPE) was used as a coupling agent for the manufacture of composites based on polyethylene (PE). The Epolene C-26 (MAPE) used here was gratefully donated by Westlake Chemicals Corporation. Yellow birch (Betula alleghaniensis) bark splinters were obtained directly after the rosser-head debarker, from the sawmill

Bois Poulin North Inc., St-Jacques de Leeds, Quebec, Canada. The bark was air dried at ambient temperature until reaching constant moisture content (9%) and stored adequately. Size separation was carried out by sieving after drying. Fibers of black spruce (Picea mariana) bark were obtained by refining at the pilot plant of FP Innovations installations in Quebec, QC, Canada. The weight and moisture of this bark were measured (above 5% moisture content) and stored properly.

PRECIPITATION OF INDUSTRIAL LIGNIN

Carbon dioxide was used to precipitate lignin by decreasing the pH of black liquor (Kruger Wayagamack and Domtar Windsor) from the initial pH 13-14 to 8-9 following the procedure inspired by LignoBoost process. A purification step consisted of washing lignins with sulfuric acid and distilled water. These precipitation and purification protocols are designed to contribute to greenhouse gas sequestration ($CO2$) and sustainable chemistry.

Esterifications of lignins with butyric anhydride

The commercial Kraft lignin (Indulin AT) was dried in an oven at 80°C overnight. Various amounts of butyric anhydride were added to the reaction system to obtain a desired substitution degree. 1, 4-dioxane and 1-methylimidazole were used as solvent and catalyst, respectively. The reaction was conducted at 60°C for two hours. The product was precipitated by adding distilled water to the solution and then washed with 0.5 M solution of $NaHCO3$ and distilled water. The esterified lignins were dried at 60°C under reduced pressure overnight.

Esterifications of lignins with stearoyl chloride

Esterification applied acid chloride rather than the anhydride and the reaction time was extended to 5 hours at a temperature of 70°C. By the end of reaction, 100 ml of hexane were added to the system and the precipitated esterified lignin was repeatedly washed with a total of 220 ml of hexane and 220 ml of distilled water. The product was finally dried at 60°C under reduced pressure overnight.

Water extraction of barks

The black spruce bark fibers (obtained from disk-refiners) and the yellow birch bark particles were each extracted with boiling distilled water (30 g oven dry bark with 1L distilled water) for two hours. The objective of extraction was to determine its impact on the properties of the bio-composites. The hypothesis was that the removal of hydrophilic extractives would increase the hydrophobicity of the bark fibers and thus improve their compatibility with hydrophobic polyethylene matrix.

Manufacturing process of bio-composites

Modified, non-modified lignins and bark fibers were previously dried at 80°C overnight. In order to study the effect of esterification on compatibility of lignins with HDPE, a comparison with a commercial coupling agent, polyethylene-graft-maleic anhydride (MAPE), was made. Blends of HDPE with lignin were prepared in various proportions by melt-processing using a twin-screw extruder (Haake TW-100). The temperature was set between 150°C (at the supply) and 165°C (at the output of the extruder) with a screw speed of 40 rpm. The amount of MAPE in the blends was selected to be similar to the lignins content. An internal mixer (Haake Buchler Rheomix) at temperature fixed at 165°C for10 minutes was used to prepare the bio-composites containing black spruce bark fibers (30%) or the yellow birch particles (50%) to prepare the HDPE-lignin bio-composites. All blends were pelletized and compression molded at 160°C for 5 minutes under 3 tons pressure.

CONCLUSION

Two Kraft lignins have been precipitated by carbon dioxide from the black liquor provided by Domtar–Windsor and Kruger-Wayagamack pulp and paper mills in Quebec. The purification protocol was confirmed by their high purity and low ash contents. Esterification reactions were performed on Indulin AT with butyric anhydride and stearoyl chloride to improve lignin compatibility with HDPE but proved to be insufficient. Stearate lignin was incorporated into HDPE matrix by melt blending. However, the tensile properties of blends were not improved by incorporation of esterified lignins only, but were possible with the standard compatibilizer (MAPE). For blends produced with HDPE –Indulin AT lignin (esterified with butyric anhydride or stearoyl chloride and non-modified) no improvements of mechanical properties of the composites have been observed, which indicated that lignins esterified as mentioned were inefficient as coupling agents. The commercial coupling agent, MAPE, was however determined to be a good compatibilizer for this kind of material. The composites made with black spruce bark fibers have been determined to have superior properties to those determined for the composites produced with yellow birch bark particles. The water extraction of the bark fibers prior to their incorporation was determined not to have significant impact on the tensile properties of the bio-composites produced with them. The pre-extraction of the bark fibers was determined not to cause any deterioration of tensile properties of the bio-composites containing them. Therefore the integration of extraction step could yield interesting co-products with high added value (pharmaceutical, cosmeceutical, nutraceutical), completing thus the biorefinery concept studied here. Other acid anhydrides will be tested for lignin esterification, the cyclic anhydrides, maleic and succinic, in order to examine thus modified lignins as coupling agents in manufacturing the HDPE-bark fibers blends and their properties.

21
Chapter

SYNTHESIS OF BIOCOMPOSITES FROM NATURAL OILS

INTRODUCTION

The polymers obtained from natural oils are biopolymers in the sense that they are generated from renewable natural sources; they are often biodegradable as well as non-toxic. Some biopolymers obtained from natural oils are flexible and rubbery. Generally, they are prepared as cross-linked copolymers. Bacterial polyesters are obtained from a large number of bacteria when subjected to metabolic stress.

Most of the common oil contains fatty acids that vary from 14 to 22 carbons in length, with 1–3 double bonds. There are some oils comprise fatty acids with other types of functionalities (e.g., epoxies, hydroxyls, cyclic groups and furanoid groups) and on a molecular level, these oils are composed of many different types of triglyceride, with numerous levels of unsaturation. The triglycerides constituting any plant oil can be chemically modified so as to become monomers or comonomers for several polymerization reactions. The fatty acid esters derived from the triglyceride vegetable oils are an attractive source of raw materials for polymer synthesis. Triglyceride oils are used in food industry as well as for the production of coatings, inks, plasticizers, lubricants and agrochemicals. Vegetable oils are excellent renewable source of raw materials for the manufacture of polyurethane components such as polyols. To use soybean oil as a monomer for the preparation of polyurethanes, it should be suitably functionalized. The transformations of the double bonds of triglycerides of oils to hydroxyls and their application in polyurethanes are the subject of many studies.

These polyurethanes from vegetable oils have high strength as well as stiffness, environmental resistance and long life and it is the main technological advantages of these oils.

SYNTHESIS OF BIOCOMPOSITES

Soybean based Biocomposites

Soybean oils are biodegradable vegetable oil and available in bulk quantity. Natural soybean oil possesses a triglyceride structure with highly unsaturated fatty acid side chains. This unsaturation in these oils provides ideal monomers for the preparation of various polymers. Polymers derived from soybean oils have been investigated. Polymers from different soybean oils show different properties, and the cross-linking density of the bulk polymers affects their thermo physical properties.

By cationic copolymerization of regular soybean oil, low saturated soybean oil or conjugated low saturated soybean oil with styrene and divinylbenzene we can prepared various copolymers. It is initiated by boron trifluoride diethyl etherate which results in polymers ranging from soft rubbers to hard thermosets, depending on the oil and the stoichiometry used. The miscibility of the initiator improved with a Norway fish oil ethyl ester. The copolymerization of soybean oil with styrene and norbornadiene or dicyclopentadiene initiated by boron trifluoride diethyl etherate synthesized heterogeneous polymeric materials with good mechanical properties and thermal stability.

MODIFIED SOYBEAN OIL POLYMERS

Triglycerides have various active sites in their structures like the double bond, the ester group, the allylic carbons, and the carbons α to the ester group. Hence there are various chemical pathways for functionalization of these triglycerides .The soybean fatty acid and poly (hydroxy alkanoate) (PHA) reacted under UV irradiation to form a cross-linked biopolyester. The esterification reaction takes place along with cross-linking via a free radical mechanism. Bacterial polyester containing olefinic groups in the side chains was prepared by feeding pseudomonas oleovorans with soybean fatty acids. The cross-linked biopolyester, obtained thermally at 60°C with benzoyl peroxide initiation having highest crosslinking density.

Fish Oil Polymers

Fish oil is available as a byproduct in the production of fishmeal. It is biodegradable. It has a triglyceride structure with a high percentage of polyunsaturated omega 3 fatty acid side chains and contain as many as 5– 6 non conjugated carbon–carbon double bonds per ester side chain. It has many applications in the production of protective coatings, lubricants, sealants, inks, animal feed and surfactants.

Corn Oil Polymers

Corn oil is one of the cheapest commercially available vegetable oils. It is used in food and livestock feed as well as in the production of ethanol. It has a triglyceride structure, with carbon–carbon double bonds per molecule in fatty acid side chains. Due to high degree of unsaturation in corn oil, it is possible to copolymerize it with other monomers. Corn and soybean oils have similar chemical structures with three fatty acid chains composed of oleic acid, linoleic acid and linolenic acid.

Tung Oil Polymers

Tung oil is one of the oldest known drying oils extracted from the seeds of the tung tree. It has numerous applications in the paint industry. Its main constituent is glyceride of elaeostearic acid with a conjugated triene structure. Due to its highly unsaturated, conjugated system it rapidly undergoes polymerization. It has been polymerized by both free radical and cationic polymerizations. Tung oil is cationically copolymerized with divinylbenzene in the presence of boron trifluoride diethyl etherate to produce hard plastics. Tung oil can also be copolymerized with styrene and divinylbenzene by thermal polymerization.

Natural Linseed Oil Polymers

Linseed oil is produced from the linseed seed. It is a fatty acid ester triglyceride, composed of about 53%linolenic acid, 18%oleic acid, 15%linoleic acid, 6% palmitic acid and 6% stearic acid. It is traditionally used as a drying oil for surface-coating applications and for making it superior drying oil in terms of film properties, different olefinic monomers, such as styrene have been copolymerized with linseed oil. Linseed oil is polymerized by cationic, thermal, free radical polymerization and by oxidative polymerization.

Castor Oil Polymers

Castor oil is a vegetable oil obtained from the castor bean of castor plant. It is a triglyceride in which approximately 90 percent of fatty acid chains are ricinoleic acid. Oleic and linoleic acids are the other significant components. Epoxidized castor oil has been used for the preparation of interpenetrating polymer networks (IPN). It is observed that the cross-linked IPNs from the epoxidized oil and adducts of tung oil with maleic anhydride had very good compatibility. The hydroxyl groups of epoxidized castor oil form hydrogen bonds with the carbonyl groups in tung oil. These hydroxyls are more reactive towards tung oil adducts than epoxidized cottonseed oil. Styrenated castor oil and linseed oil can be prepared by the macromer technique. The copolymerization of dehydrated castor oil with styrene has been reported.

A blend of dehydrated castor oil and epoxy resin and the miscibility of the blends of epoxidized dehydrated castor oil and poly (methyl methacrylate) has been reported.

Polymers from Other Oils

Drying and semidrying oils such as sesame, sunflower, safflower, walnut oil are also used for the preparation of polymer by different methods. A substantial research has been done on the polymerization of high oleic sunflower oil with different comonomers. The copolymers of sunflower oil and styrene were formed via free radical mechanism. The Ritter reaction is used to study acrylamide functionality on the triglyceride of sunflower oil. Biobased resins are prepared by the addition of a derivatized vegetable oil (Methyl ester of soybean oil (MESO), epoxidized methyl linseedate (EML)) to the unsaturated polyester resin (UPE) matrix. The initiator used was methyl ethyl ketone peroxide (MEKP) and promoter used was cobalt (II) naphtenate (CoNap). This bioresins are then used for the fabrication of a biocomposites using an optimum amount of natural fibre (30% volume). The natural fibre used is a non-woven hemp mat, which contains 10% poly (ethylene terephthalate) (PET) as filler. It produced plastic and composite plaques which were cut into required shapes for various tests.

Hydro chlorofluorocarbon (HCFC) and pentane blown rigid polyurethane (PU) foams are synthesized from polyols derived from soybean oil. The foams prepared from these polyols were found to have comparable mechanical and insulating properties to those of commercially available polypropylene oxide (PPO)-based foams. It was observed that the soybean polyol derived PU foams were more stable.

Researchers also studied the thermal stabilities of PU-based on various vegetable oils, the structure and properties of PUs prepared from halogenated as well as non-halogenated soybean polyols.

They modified epoxidized soybean oil (ESBO) with hydrochloric acid, hydrobromic acid, methanol and hydrogen. From these polyols, four types of polyol polyurethanes were prepared. A series of urethane acrylates can be derived from soybean fatty acid modified hyper branched aliphatic polyesters. Nahar (Mesua ferrea L.) seed oil contains mainly triglycerides of linoleic, oleic, palmitic and stearic acids. Polyester, polyesteramide, polyurethane and polyurethane amide resins are synthesized from nahar seed oil. Synthesized aliphatic amine, and phthalic anhydride can be employed as crosslinking agents for the transesterfication and epoxidation of regular food grade soybean oil to produce Epoxidized allyl soyate (EAS), novel soy based epoxy resin. Soy-based resin can be produced by transesterfication and epoxidation of regular food grade soybean oil. Radical copolymerization of the soybean oil monoglyceride maleates with styrene yield thermosetting liquid molding resins whereas sheet moulding compound resins can be derived from soybean oil by using maleated hydroxylated soybean oil (MHSO) and maleated acrylated epoxidized soybean oil (MAESO) with styrene, MgO as thickening agent, Formic acid for the epoxidation reaction.

Polyurethane (PU) flexible foams can be synthesized by substituting a portion of base polyether polyol with soybean oil-derived polyol (SBOP) and substituent, crosslinker polyol and styrene acrylonitrile (SAN) copolymer-filled polyol. Biopolymer from the palm oil resources can be synthesized.

CONCLUSION

In recent years, natural oils have become the best alternative for petroleum-based polymers. The fossil based monomers are harmful to environment. These are non-renewable as they are derived mainly from petroleum-based materials and day by day fossil-based feed stocks are depleting very rapidly. Hence, it has become the main goal for the researchers in coming years to produce viable polymers from the natural resources. Different research groups in the world are studying the properties of natural oils and their composites for utilization as polymers, resins, varnishes and paints.

The vegetable oils provide a large variety of options for the preparation of different polymers. All the vegetable oils are triglycerides of fatty acids and most contain unsaturated groups. Only a few contain other groups such as hydroxyl in castor oil, safflower oil and lesquerella oil, and oxirane group in vernonia oil. The unsaturation present in these oils makes them ideal for the preparation of bio-based polymers. The polymerization of these oils is carried out via free radical and cationic polymerization reactions

22
Chapter

NANO-LIGNIN FILLED NATURAL RUBBER COMPOSITES

INTRODUCTION

Lignin, the second most abundant renewable natural resource next to cellulose, is a highly-branched, three dimensional biopolymer. It consists of three phenylpropanoid units, such as p-hydroxyphenyl (H), guaiacyl (G) and syringyl (S), which are attached to one another by a series of characteristic linkages (β-O-4, β-5, β-β,) etc. The major chemical functional groups in lignin include hydroxyl, methoxyl, carbonyl and carboxyl groups in various amounts and proportions, depending on genetic origin and applied extraction processes.

The distinct network structure as well as the presence of the various chemical substituents confers special functional properties to lignin, such as stabilizing effect, reinforcing effect, UV-absorption, biodegradability, anti-fungal and antibiotic activity. However, the potential of lignin is not clearly valued, because it predominantly obtained as byproducts in pulp production is mainly used as fuel. Fortunately, incorporating into polymeric materials will be a value-added application for lignin. Especially in rubber-based materials, lignin is less dense, non-conducting, and being lighter in color compared to carbon black. It appears more amenable for the preparation of light-colored rubber compounds.

So, for the past decades great amounts of researches and patents of lignin as filler added to rubber compounds have investigated carbon black replacement in order to achieve similar reinforcement of rubber composites. Usually, two strategies (dry-milling and co-precipitation) are mostly used for the preparation of lignin-filled rubber composites. Lignin as dry powder straightforward milled into rubber shows almost no reinforcing effect. This is believed to be a result of thelignin particles adhering together by intermolecular hydrogen bonding and thus not being dispersed into the rubber by milling . It is a unique property that lignin is soluble in aqueous alkali because of the ionization of the phenolic hydroxyl and carboxylic groups, such solutions being compatible with rubber latex in all proportions. Furthermore, the co-precipitation of the lignin and the rubber from the mixing solution with acids is possible in the same pH range. In comparison to dry-milling method, the distribution of the lignin in the rubber compounds prepared by co-precipitation method is more homogeneous and the particle size is finer.

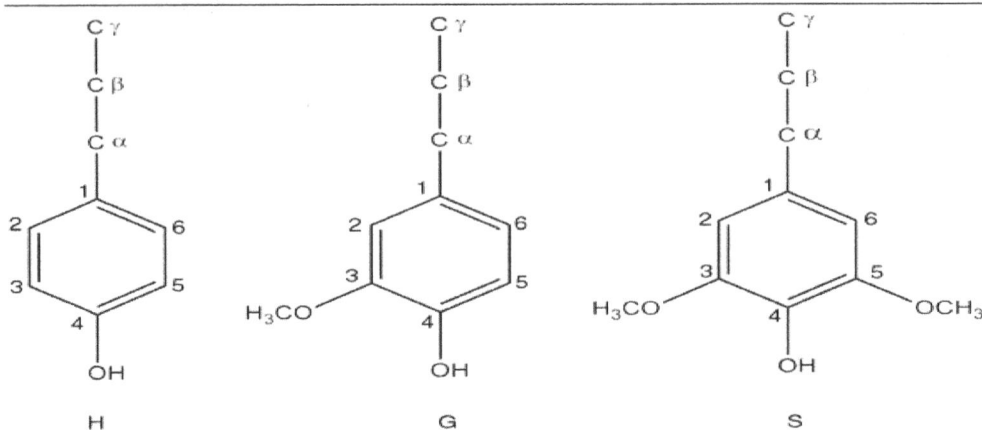

Figure 1. Three phenylpropanoid units of lignin

Even so, the lignin is still unable to be dispersed at nanoscale. Nevertheless, the reinforcing effect of lignin for rubber compounds intensively depends on particle size and strong interfacial bonding with rubber matrix. In most reports, the lignin shows little reinforcing effect on nonpolar rubber matrix and even deteriorates the performance of rubber composites. To the best of our knowledge, there is no report on nanoscale lignin reinforced natural rubber. Therefore, we highlight the fabrication of nano-lignin and the nanoscale distribution of lignin in rubber matrix by a novel strategy.

Lignin can be viewed as an anionic polyelectrolyte when phenolic hydroxyls and carboxylic groups are ionized. Usually, three types of polyelectrolyte complexes can form: soluble, colloidal and coacervate. In previous publications, the studies on the formation of polyelectrolyte complexes between cationic polyelectrolytes and anionic lignin aim to improve wastewater-treatment efficiency, increase the efficiency of retention aids and enhance the strength of papers in papermaking. As a consequence, the lignin-cationic polyelectrolyte complexes are commonly coacervates for those purposes.

However, the complexes must be colloidal to mix with natural rubber latex (NRL) and produce NR/lignin nanocomposites. In the present work, the formation and characteristics of colloidal LPCs were investigated and the adsorption mechanism of PDADMAC onto the lignin surface was discussed. The morphology of the NR/lignin composites was also observed by SEM. The mechanical performance and vulcanization behaviors were studied, as well as the thermal and thermo-oxidative stability.

EXPERIMENTAL SECTION

Materials

Lignin was industrial sulfate lignin (micron scale), purchased from Shandong Tralin Pape Co., Ltd., China. The average molecular weight of sulfate lignin was 3801 with the polydispersity index of 2.15. PDADMAC (100 000–200 000; 20 wt% in water) was obtained from Sigma-Aldrich (Sigma- Aldrich, Louis, MO). NRL stabilized by ammonia with a total solid content of 64.5% was provided by the Tropical Crops Research Center of Zhanjiang, China. Sodium hydroxide, hydrochloric acid and sulfuric acid were analytical grade, purchased from Guangzhou Chemical Reagent Factory, China. The reagents used in rubber formula such as sulphur, stearic acid, zinc oxide, N-tert-butyl-2-benzothiazole sulfonamide (CBS), were industrial grade, kindly provided by South China Rubber Tyre co., Ltd.

Purification of the industrial sulfate lignin

The industrial sulfate lignin was dissolved in a beaker with deionized water followed by adding NaOH to adjust the pH to approximately 13. The resulting solution was centrifuged to exclude fibril or insoluble impurities before slowly titrating with 1.0 N sulfuric acid to a final pH of approximately 2 with mechanical stirring. Subsequently, the slurry was placed in water bath at 70°C for 2 h to accelerate lignin agglomeration then filtered. The lignin was washed with deionized water three times and dried in a vacuum oven at 50°C for 24 h followed by extraction with pentane to remove the organic impurities. Finally, the purified lignin was dried at 50°C and stored under reduced pressure.

Preparation of sample solution

Purified sulfate lignin was dissolved in deionized water at a mass concentration of 0.5% and the pH was adjusted to 12. PDADMAC aqueous solution was diluted to 2 wt % with deionized water followed by adjusting the pH to 12. The various mass of sulfate lignin solution was dropped slowly into the PDADMAC solution with vigorous stirring, leading to formation of colloidal LPCs with different mass ratios of lignin to PDADMAC.

Preparation of NR/lignin compounds

The NRL was diluted to a solid content of 10% and its pH was adjusted to 12 with 1 N NaOH. After that, a desired amount of the lignin solution or the colloidal LPCs solution was then dropped into NRL under vigorous mechanical stirring. The resulting solution was co-precipitated by adjusting the pH to 2 with 1 N sulfuric acid. Finally, the obtained mixtures were filtered, water washed and dried in a vacuum oven at 50°C. All the rubber compositions were summarized in Table 1. The dried composites and other additives were mixed on an open two-roll mill.

Then the compounds were vulcanized in a standard mold at 143±1°C for optimum vulcanization time, which was determined by the U-CAN UR-2030 vulcameter (Taipei, Taiwan).

Table 1. Recipe of NR/lignin compounds

Ingredients [phr]b	Sample code								
	L-0	L-1	L-3	L-5	L-7	LPCs-1	LPCs-3	LPCs-5	LPCs-7
NR	100	100	100	100	100	100	100	100	100
Lignin	0	1	3	5	7	0	0	0	0
LPCSc	0	0	0	0	0	1	3	5	7

* Rubber ingredients: ZnO 5, Stearic Acid 2, CBS 2, S2.
* PHR, parts per hundred of rubber byweight.
* The ratio of lingnin to PDADMAC is 0.45, and the amount of LPCs in the formula refers to the weight of lignin in the LPCs.

Characteristics of colloidal LPCs

It is generally known that oppositely charged polyelectrolytes form complexes over a broad range of stoichiometric ratios and that the complexes tend to be water-soluble unless they are nearly stoichiometric. In this work, the colloidal LPCs were fabricated by drop wise adding the alkali solution of lignin into the PDADMAC solution. Hence, the formed LPCs were positively charged and water-soluble until the mass ratio of lignin to PDADMAC exceeded stoichiometric point.

However, the NRL particles were natively charged at the pH of 12, as the protein molecules absorbed on the surface of NRL particles contained carboxylic and amino groups and the carboxylic groups would be ionized at that pH. When the LPCs solution were added into NRL, the natively charged NRL were subsequently adsorbed onto the positively charged LPCs via electrostatic self-assembly, which would suppress the aggregation of lignin and finally resulted in homogeneous distribution of lignin in NR matrix.

However, excess PDADMAC adsorbed onto lignin is unnecessary. If present in excess, it will prematurely flocculate NRL and reduce the loading of lignin incorporated into rubber matrix. To optimize the fabrication of LPCs, the effect of mass ratio of lignin to PDADMAC on particle size of LPCs was investigated. The particle size of LPCs is consistently reduced (from about 400 to about 180 nm) with the increasing lignin/PDADMAC ratio (from 0.05 to 0.5). It should be noted that the minimum particle size of LPCs is still larger than that of the collapsed lignin at the pH 4 (about 110 nm). The particle size should be the size of collapsed lignin plus the thickness of absorption layer of PDADMAC onto lignin surface. A significantly increasing particle size follows when lignin and positively charged LPCs are present.

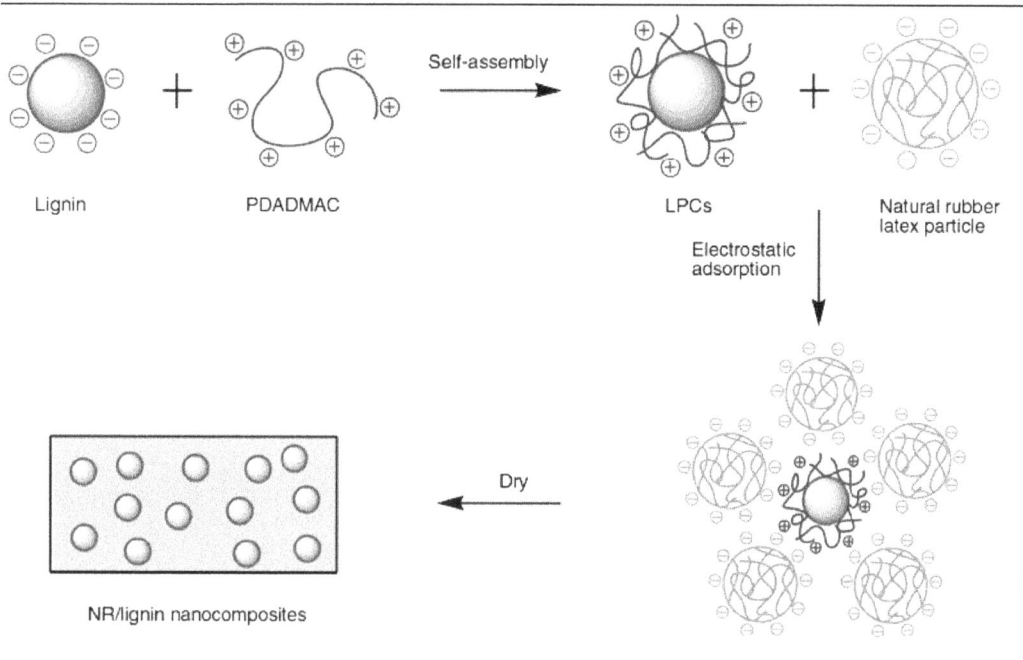

Figure 2. Schematic illustration of process for NR/Lignin nanocomposites

CONCLUSIONS

Natural rubber/lignin nanocomposites were successfully prepared via co-precipitation of colloidal lignin-cationic polyelectrolyte complexes and rubber latex. PDADMAC was adsorbed onto lignin particles through ion-ion interactions, cation-! and!-! interactions which were confirmed by UV-vis and FTIR. As employing proper amount of PDADMAC, nanoscale LPCs were formed, laying a foundation for fabrication NR/LPCs nanocomposites. LPCs accelerated the vulcanization of NR/LPCs nanocomposites.

Furthermore, LPCs were homogenously distributed in NR matrix, which resulted in improved mechanical properties, thermal and thermo-oxidative stability of NR/LPCs composites. In a word, this paper presents a promising strategy to utilize lignin for partial carbon black replacement.

23
Chapter

HYBRID BIO-BASED POLYMERS FROM BLENDS OF EPOXY AND SOYBEAN OIL RESINS REINFORCED WITH JUTE WOVEN FABRICS

INTRODUCTION

The usage of natural materials in the production of composite structures is increasing day by day, with the increasing knowledge about environmental factors. The term of bio-composite is often used for the composites which have partially natural components; either a matrix or reinforcement material. However, it is essential to provide a combination in which both components (matrix and reinforcement material) are natural in order to truly entitle it as a bio-composite.

Bio-composites, which have a wide range of usage, are generally composed of petroleum-based thermoset or thermoplastic resins reinforced with natural textile fibers. Although the thermoplastic resins obtained from renewable sources are commercially available, thermoset resins are mostly produced from crude oil.

Environmental damage, climate changes, and diminishing fossil sources have led firms and scientists to search for alternative matrix materials and thus bio-based plastics have emerged. The development of bio-based polymers from cheaper and renewable materials instead of petroleum-based materials has significant impacts on the economy and the environment. In addition to its many advantageous properties, the most important feature of bio-based polymers is their overall intrinsic low toxicity.

Petroleum-based epoxy resin is the most widely used thermoset polymer in composite materials due to its low cost, ease of use and high mechanical properties. However, the uncertainties about petroleum supplies, prices and their effects on the environment reveal the fact that the use of epoxy needs to be reduced and substituting biomaterials are required.

Considering the automotive sector, which is one of the sectors where composite materials are used most, it is seen that the most important thing to increase the efficiency of automobiles and reduce fuel consumption to diminish its hazardous effects to the environment is to reduce the weight of the vehicle. Studies in the literature and the prototypes produced have shown that a weight gain of 25% can be achieved by using bio-resin based composites instead of steel. In addition to being lightweight, composites offer ease of production and workability compared to metals.

Thermoset bio-resins obtained from plant oils (soybean oil, flaxseed oil, linseed oil, palm oil, rapeseed oil, etc) which are unsaturated triglycerides are promising raw materials for the 'greening' of thermosets.

Unlike synthetic resins, they are derived from plants and contain carboxylic acid, oil, and isoprene-based hydrocarbons. Bio-resins do not show toxic properties during production and can be used in admixture with all synthetic resins. They are also odorless and have dimensional stability.

Among the existing plant oils, soybean oil seems to be the most attractive alternative resource due to its very low price and abundant supply. The building block of soybean oil resin is a triglyceride, the main component of vegetable and animal fats. Epoxidized soybean oil (ESO) is known as a common plasticizer for polyvinyl chloride. AESO is commercially available soybean oil produced from the epoxidation of soybean oil followed by reaction with acrylic acid. AESO shows non-volatile and non-toxic behaviors. Typical soybean oil is a triglyceride with~4.5 unconjugated C=Cbonds. In addition, AESO is highly viscous at room temperature and has a low crosslinking capacity due to its long aliphatic chains and a low degree of unsaturation.

On the other hand, the variety of natural reinforcement materials (natural textile fibers, hazelnuts, pine needles, sawdust, etc) is increased day by day, but natural resin is a new study topic for researchers. Natural fiber reinforced composites have many advantages over synthetic fiber reinforced composites. Some of these are less damage to nature, more fiber/less polluting polymer content to achieve the same performance characteristics and their end of life is result in recovered energy and carbon credits. In addition to these advantages, the use of natural fiber-reinforced composites also has positive effects on economic issues.

Best fibers are gathered from certain natural plants' skin or inner shell of the stem. They have superior mechanical properties (such as high strength and rigidity) than other plant fibers and due to this fact; they are the most used natural fiber reinforcement materials. Among the best fibers, jute is the most preferred fiber in composite production. It has properties such as; high modulus, low cost, good thermal and electrical insulation abilities, moderate moisture regain, biodegradability, silky luster, high specific strength and commercially availability While the matrix material provides the surface appearance and the resistance to environmental factors in general, the reinforcing material ensures the integrity of the composite material by holding the fibers together in addition to strength and load-bearing properties.

In the literature, the studies related to the AESO based composites mostly focus on the use of AESO in a mixture of several matrix types such as styrene, polyester, vinyl ester, epoxy, etc. There are only a few studies in the literature that works on natural textile fiber reinforced composites consisting of soybean oil resin and epoxy resin in order to improve the toughness of the neat epoxy composites. In one of the studies, Bakar et al mixed epoxidized soybean oil with petroleum-based epoxy resin in different ratios to investigate its effects on mechanical (tensile, impact and flexural strength) and thermal properties.

Table 1. Density and viscosity values of matrix system elements

Matrix components	Density	Viscosity (mPA.s)
Epoxy	1.15	1200-1400
Hardener	1.00	10-20
Soybean Oil, Acrylated expoxidized	1.04	18000-32000

Results indicated that although a slight increase of mechanical properties was observed at 10% epoxidized soybean oil content sample, tensile, flexural and thermal properties of composite structures decreased with ascending epoxidized soybean oil content. When the impact strength results were examined, it was seen that epoxidized soybean oil content enhanced the impact properties of the samples owing to its plasticization effect.

Sahoo et al used sisal fiber as reinforcement material in epoxidized soybean oil/epoxy blend resin to enhance the mechanical and thermo-mechanical properties. With the addition of sisal fibers, tensile strength values were increased. While with increasing soybean oil ratio, tensile strength values were decreased and elongations at break values were increased. In one of the studies, Kocaman and Ahmetli fabricated banana bark and seashell reinforced epoxy/AESO (50 wt%/50 wt%) based composites to examine the effects of those natural fillers on the tensile strength, elongation at break and hardness values of composites. As for the impact of AESO, AESO improved the plasticity of the epoxy resin. In another study, Temmink et al reinforced bio-epoxy and AESO resin with post-consumer denim waste with four different manufacturing techniques (compression molding, vacuum infusion, resin transfer molding and hand lay-up). The results showed that bio-epoxy composites have superior properties compared to AESO composites, while both bioepoxy and AESO-based composite materials are found suitable for structural applications. In a study of Niedermann, the impact of the addition of epoxidized soybean oil on mechanical, storage modulus and glass transition temperature of jute fabric reinforced epoxy composites was investigated. Three types of epoxy resins (aromatic diglycidylether of bisphenol-A (DGEBA) resin, a glycerol (GER)-and a pentaerythritol (PER)- based aliphatic resins) were utilized as a base resin. From the results, it was revealed that while increasing epoxidized soybean oil content in aliphatic composites increased the Tg value, it decreased Tg in the aromatic system. In all systems, the ascending amount of epoxidized soybean oil content decreased the mechanical properties. However, while the mechanical properties obtained in aromatic systems were higher than those of pure epoxidized soybean oil composites when the epoxidized soybean oil ratio in aliphatic systems exceeded a certain value, the mechanical properties reached a lower level than those of pure epoxidized soybean oil composites. Liu et al utilized two types of reinforcement materials (bamboo and hemp fibers) and 4 different matrix materials (AESO, AESO with methacrylated isosorbide (MI) as a comonomer (MI-AESO), methacrylated AESO with MI (MI-MAESO) and MI resins) to analyze the synergistic effects of fiber type and resin composition on properties of bio-composites. Results indicated that hemp fiber reinforced composites had higher flexural strength, moduli and water resistance compared to bamboo fiber reinforced composites.

Moreover, MI-AESO, MI-MAESO, and MI-based composites had higher mechanical strength, modulus and glass transition temperature compared to AESO based composite because of its' flexible structure.

In this study, AESO modified epoxy resin reinforced with four-ply jute woven fabrics and are produced by the vacuum infusion method. When literature is examined, it is seen that mostly four, six, eight and ten plied fabric reinforced composite plates were produced. Since the main aim of this study is to investigate the effect of bio-resin ratio on the properties of composite material, the minimum number of fabric layers is chosen. The optimizing the resin content in the composite structure is one of the most important parameters in composite production. The reinforcement material should be saturated with matrix material with as little excess as possible. The technique of 'squeezing out' the excess resin to maximize fiber-to-resin content is the main theory of the vacuum infusion system. The fiber/matrix ratio, tensile strength, flexural strength, drop weight impact resistance and Charpy impact strength together with differential scanning calorimetric analysis, Fourier-transform infrared spectroscopic analysis and scanning electron microscopy analysis of the hybrid biocomposites are examined in order to determine the effect of AESO blends in physical, mechanical, instrumental and morphological properties of the composites.

MATERIALS AND METHOD

Materials

Jute woven fabric (300 gm-2) is used as a reinforcement material in bio-composite production, while epoxy resin (F-1564, Fibermak), hardener (F-3486, Fibermak) and acrylated epoxidized soybean oil (AESO, Sigma Aldrich) are used as matrix material. The density and viscosity values of matrix system elements are given in table 1.

METHODS

Composite fabrication

The composite structures within the scope of this study are produced by the vacuum-infusion method under 1 bar pressure at room temperature (20°C ± 2°C). The applied vacuum helps to distribute the homogeneous resin flow from the jute fabrics, while vacuuming the excess amount of resin in the composites, thereby achieving constant composite thickness. The vacuum infusion set-up is shown in figure 1(a). For the performance analysis, samples are cut by CNC milling machine at 0° and 90° directions of composite plates (figure 1(b)).

For each composite sample, four-ply jute woven fabric is used as reinforcement material. The proportions of the AESO polymer in the resin system range from 0 to 100 wt% by weight in the increment of 10%. When preparing resin systems, 25 wt% of hardener is used in the whole resin system. The remaining 75% is used for predetermined proportions of AESO resin, epoxy resin or mixtures thereof. As an example of the sample code,

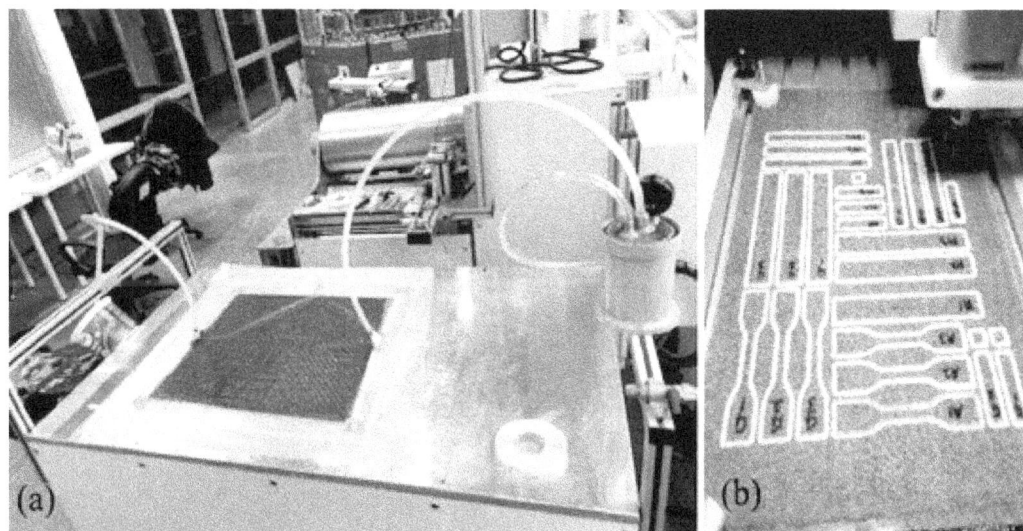

Figure 1. (a) The vacuum infusion set-up, (b)Cutting test samples by CNC milling machine.

Table 2. Sample codes of composite designs.

Sample code	Epoxy ratio (%)	AESO ratio (%)	Hardener ratio (%)
0BIO	75	-	25
10BIO	57.5	7.5	25
20BIO	60	15	25
30BIO	52.5	22.5	25
40BIO	45	30	25
50BIO	37.5	37.5	25
60BIO	30	45	25
70BIO	22.5	52.5	25
80BIO	15	60	25
90BIO	7.5	67.5	25
100BIO	-	75	25

0BIO indicates 0 wt% AESO, 75 wt% epoxy and 25 wt% hardener, while 10BIO indicates 7.5 wt% AESO, 67.5 wt% epoxy and 25 wt% hardener. Hardener ratio is kept constant for all resin systems. In this encoding type, samples are encoded in the range 0BIO to 100BIO. The codes of the samples according to the resin system ratios are given in table 2.

FTIR analysis

The FTIR results applied to the resins (AESO and epoxy) and reinforcement material used in the project are given in figures 2–4.

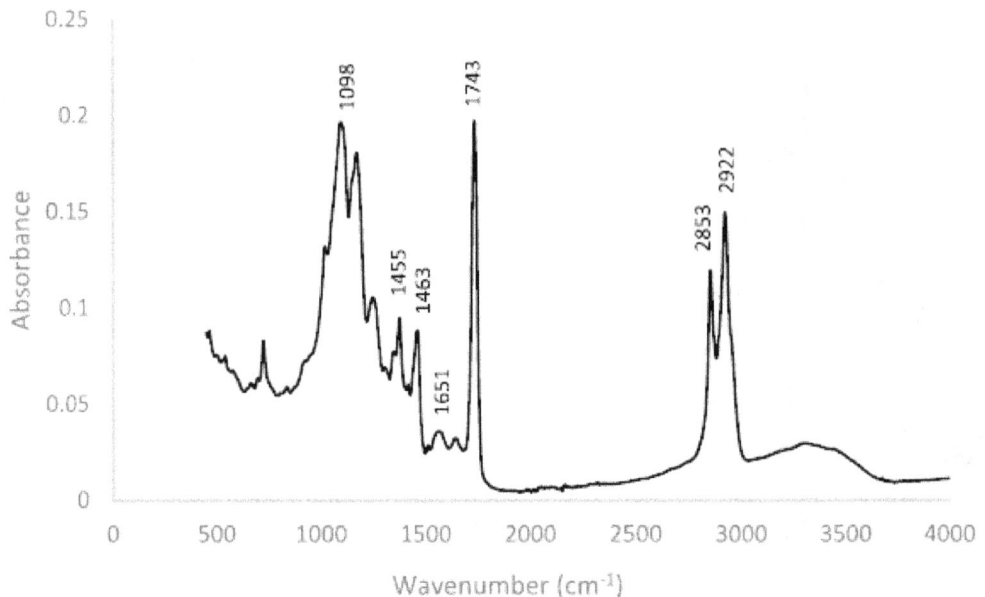

Figure 2. FTIR analysis of AESO resin.

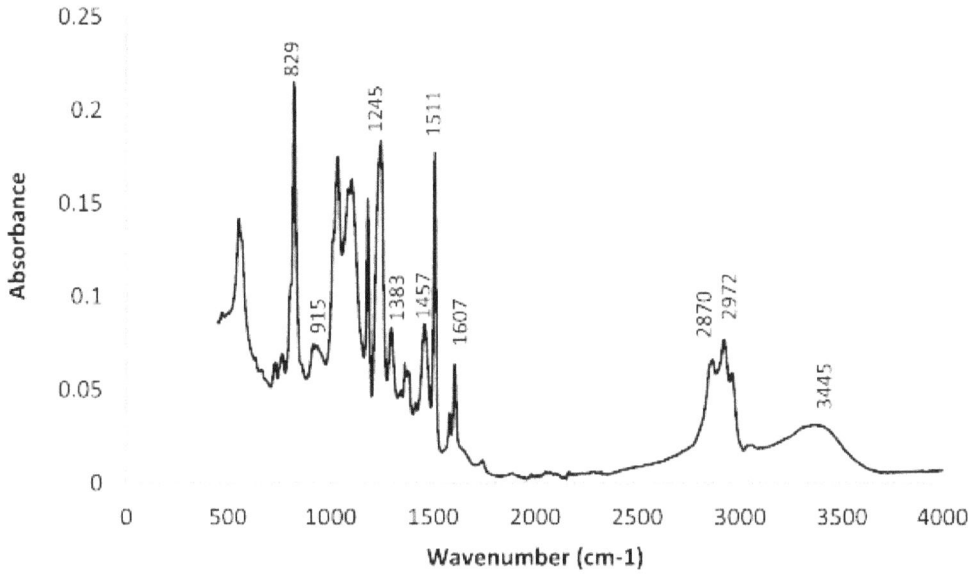

Figure 3. FTIR analysis of epoxy resin.

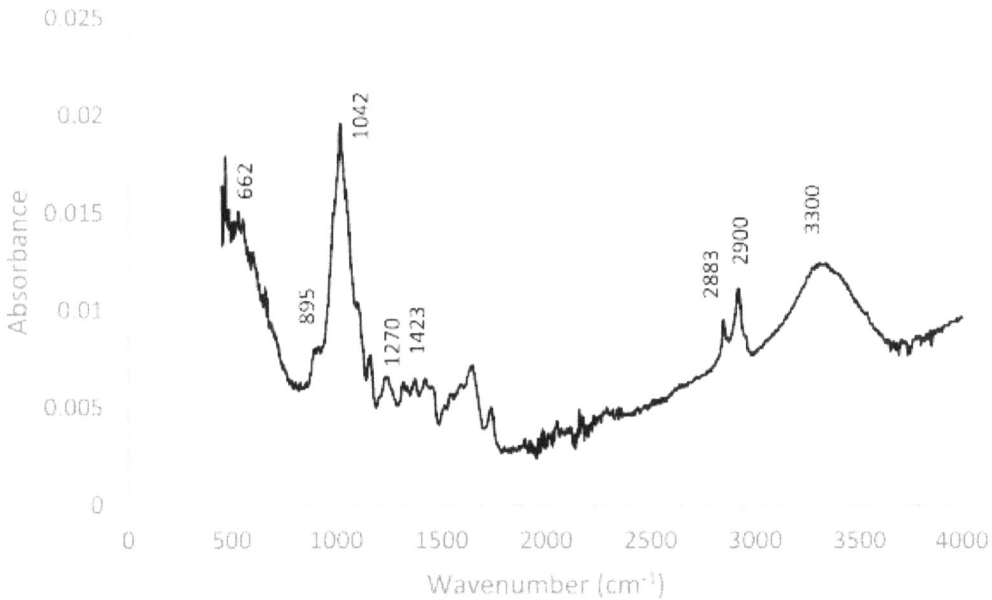

Figure 4. FTIR analysis of jute fabric.

CONCLUSION

Considering the life cycles of environmentally friendly bio-composites, it is obvious that they will contribute to the national economy in the sense of solid waste management in the long term.

24
Chapter

BIPOLYMERS FROM ROSIN-BASED CHEMICALS

INTRODUCTION

The natural origin, low price, abundance and chemical modification potential of rosin make it a valuable raw material in numerous applications. Besides the mentioned advantages, rosin is also safe for living organisms. Its derivatives are claimed as non-toxic as well, despite their allergenicity. This unique set of beneficial properties of rosin determines it as an attractive subject of innovative research characterized by a considerably growing interest in recent decades.

Unfortunately, the awareness of the possibility of using rosin as a raw material for obtaining chemicals is generally unsatisfactory. There is a burning need to bring this topic to the attention of a larger group of scientists. Furthermore, the growing number of publications causes difficulties in keeping up with the latest research, as well as in selection of more promising discoveries. Sadly, the last comprehensive review document dedicated exclusively to rosin and its modifications was published in 2008. Since then, only fragmentary information on rosin has appeared in review articles on bio-based polymers and resin systems, as well as in reviews on rosin derivatives in catalysis, controlled drug-delivery systems and small-molecule compounds.

ROSIN-BASED CHEMICALS

1. General Comments

It includes preparations of completely new chemicals, as well as new ways to synthesize already known compounds. Only products with declared or obvious practical applications have been chosen. The review is divided into two sections according to the general molecular structure of the prepared compounds, while each section is divided according to practical application and the structure similarity of the compounds. The collected data are given in the article text and in schemes. The following information is presented in the text: product name, product morphology, substrates name(s), separation techniques and practical applications. On the other hand, data such as reaction scheme, catalyst use, reaction media, temperature, pressure, time and yield are included in the schemes. The more data provided, the more advanced the research on a compound, and the more reliable the recipe. The lack of data should indicate that the research on a certain rosin derivative is probably at an early, basic stage. It has to be underlined that reaction schemes do not take into account advanced stereochemistry. Finally, almost all reactions were conducted under an inert atmosphere: nitrogen or argon, so reaction schemes do not include this information.

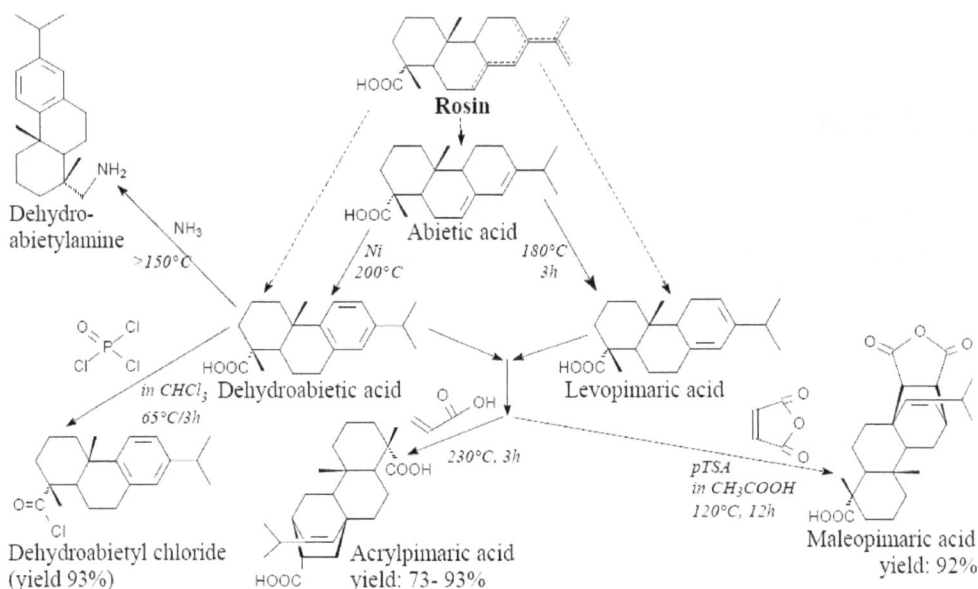

Scheme 1. Synthesis of essential rosin-based substrates/intermediates

2. Small and Medium Molecule Compounds

This section describes rosin-derived small/medium molecule compounds with strictly defined structures that do not contain the repeated units typical for macromolecules.

INTERMEDIATES

Intermediates are the products of simple rosin modifications, which are necessary for the preparation of many new compounds. They can be synthesized in very simple, well described ways with high yields. The sustainability of these processes is high: They usually use a biobased main substrate (rosin) in solvent-free processes. Some drawback can be the separation processes that may not always be easy, because of the high viscosity, m.p. and stickiness of products. In view of the above, rosin-based intermediates have very high commercialization potential and some of them are commercially available in certain regions of the world.

Maleopimaric acid is an off-white solid (m.p. 223°C). It can be prepared from abietic acid and maleic anhydride via a Diels-Alder reaction according to Scheme 1, followed by recrystallization. Maleopimaric acid is one of the crucial products in rosin chemistry. It can be used directly as an epoxy resin hardener, but its applications are much wider. They include preparation of epoxy resins, acrylic resins, allyl resins, polyols, bio-based curing agents for synthetic epoxy resins or bio-based ones, surfactants, intermediates, biologically active compounds, polyurethanes, chemicals for NMR techniques and photo litography, sorbents, organo silicon compounds, printing inks and for the hydrophobization of wood surfaces. It is noteworthy that the main degradation products of maleopimaric acid are water, carbon dioxide, formamide and also aliphatic and aromatic derivatives. It is worth noting that fumaropimaric acid is an isomer of hydrated maleopimaric acid, which can be synthesized in a similar way, but its importance to rosin chemistry is much smaller: it can be used in synthesis of triglycidyl epoxy resin and water-borne polyurethanes.

Acrylpimaric acid is an off-white solid (m.p. 220°C). It can be prepared from acrylic acid according to Scheme 1, followed by precipitation, filtration, washing and recrystallization. It is noteworthy that neat rosin can be acrylated as well. As an important material in rosin chemistry, acrylpimaric acid can be used for preparation of diallyl acrylpimarate, polyesters of acrylated rosin and polyethylene glycols, epoxy resins, acrylpimaryl dichloride, acrylpimaric acid amides, cyclic diamide, quaternary ammonium salts, calcium and zinc salts as well as polyesters. Moreover, it can be used directly as an epoxy curing agent.

Another important compound in this subsection is dehydroabietyl chloride. It is a viscous, yellow, oily liquid. It can be prepared using: (i) oxalyl chloride according to Scheme 1 prior to evaporation of unnecessary substances, (ii) phosphorus trichloride in chloroform (yield 92.5%) or (iii) thionyl chloride in presence of 4-dimethylaminopyridine. It can be used as a substrate for the synthesis of macroinitiators for atom transfer radical polymerization (ATRP) reactions, rosin phosphate esters, N-hydroxyethylacrylamide ester of dehydroabietic acid, dehydroabietic ethyl methacrylate, dehydroabietic propargyl ester dehydroabietic hexyl acrylate, as well as other intermediates in synthesis of various surfactants and medicines.

Dehydroabietylamine, also known as leelamine, is a solid (m.p. 44.5°C). It is commercially available. Its application in antitumor therapies was investigated in recent years. Moreover, it can be a substrate for preparation of epoxy resin, bio-based benzoxazines, quaternary ammonium surfactants, acrylic monomers: glycidyl methacrylate monomer or Ndehydroabietic acrylamide.

Resins and Monomers

Rosin-based resins and monomers are compounds which contain epoxy, acrylic, allyl, hydroxyl or oxazine reactive groups that enable cross-linking, polymerization or building in the polymer matrix. Their preparations are usually well described and easy to perform. The syntheses are similar to conventional resins/monomers preparations; however the necessity of using organic solvents in several reactions is a disadvantage in the context of Green Chemistry rules. The high modification potential of rosin derivatives allows one to prepare resins and monomers showing diverse and designable properties. They can exhibit adjustable glass transition temperatures, low volume shrinkage, as well as improve elastic modulus, Young's modulus, shape-memory, flame retardancy, corrosion protection features of final casts/polymers in comparison with petroleum-based compounds. Therefore, the best described recipes show high commercialization potential in the segment of resins, adhesives and paints, but in most cases, additional applied research should be performed to increase their technology readiness level (TRL).

Triglycidyl ester of maleopimaric acid is beige, viscous liquid. It can be prepared from maleopimaric acid, epichlorohydrin and sodium hydroxide according to Scheme 2, followed by filtration, washing and evaporation. It can be used in liquid epoxy resins, non-cytotoxic biobased epoxypolyurethanes as well as synthesis of rosin-based cyclic carbonates. Similar liquid resin can be synthesized from fumaropimaric acid.

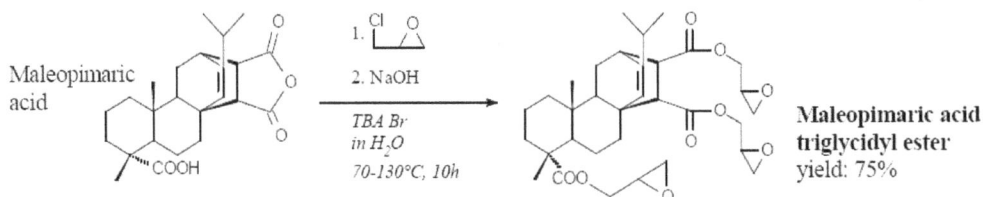

Scheme 2. Synthesis of triglycidyl ester of maleopimaric acid.

Diglycidyl acrylpimarate is a yellowish liquid. It can be prepared from acrylpimaric acid, epichlorohydrin and sodium hydroxide according to Scheme 3, prior to filtration, washing and drying. It can be used in epoxy materials showing improved thermal, mechanical and shape memory properties.

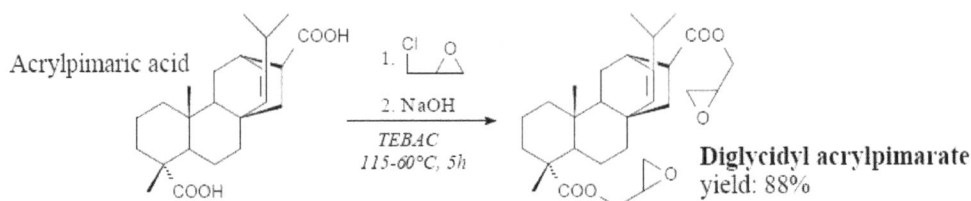

Scheme 3. Synthesis of diglycidyl acrylpimarate.

Another diglycidyl derivative of acrylpimaric acid and its siloxane modification can be prepared according to Scheme 4, prior to washing, filtration and vacuum evaporation. Prepared epoxy resins can improve the thermal stability and flame retardancy of products.

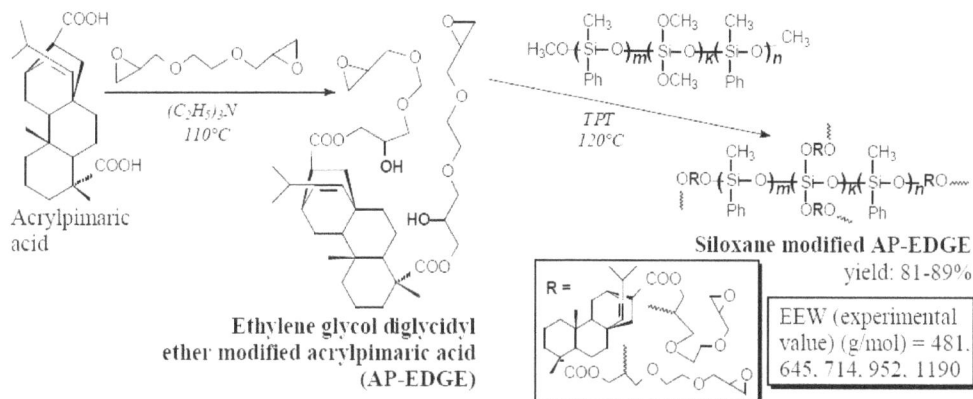

Scheme 4. Preparation of ethylene glycol diglycidyl ether modified acrylpimaric acid and its siloxane derivative.

Dimaleopimaryl ketone is a brownish yellow solid. It can be prepared from levopimaric acid and maleic anhydride according to Scheme 5, prior to washing and recrystallization. It can be used directly as an epoxy resin hardener, as well as for the synthesis of bio-based epoxy resins, i.e. tetraglycidyl dimaleopimaryl ketone, according to Scheme 5.

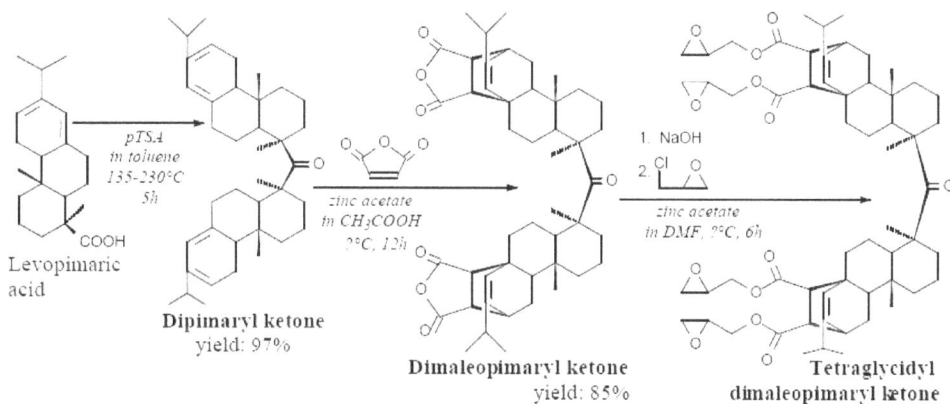

Scheme 5. Synthesis of tetraglycidyl dimaleopimaryl ketone.

Rosin pentaglycidyl ether is a solid. It can be prepared from epoxidized rosin, water, potassium hydroxide and epichlorohydrin according to Scheme 6, and using such separation methods as vacuum drying, filtration, precipitation and washing. It can be used as a component in epoxy resin systems showing high glass transition temperature as well as elastic modulus.

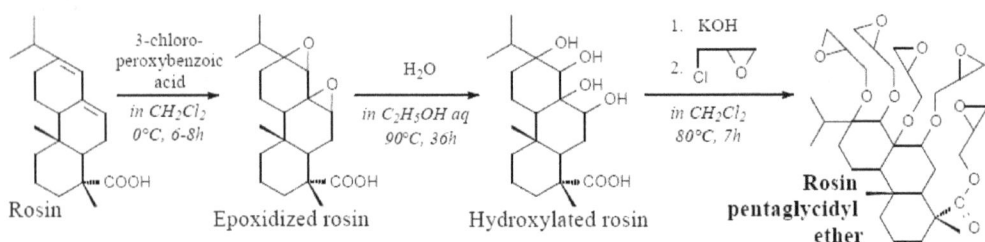

Scheme 6. Synthesis of rosin pentaglycidyl ether.

Polygral is a solid byproduct of the paper and forestry industry, containing rosin acids and their oligomers. It can be epoxidized to prepare bio-based epoxy resins. Endocyclic epoxidized polygral is a red brown solid, which can be prepared using 3-chloroperoxybenzoic acid, according to Scheme 7, prior to vacuum evaporation. On the other hand, exocyclic epoxidized polygral is a viscous red brown liquid, that can be prepared in two ways, using oxalyl chloride or N,N'-diisopropyl carbodiimide before addition of glycidol, according to Scheme 7 prior to washing and vacuum drying. They can be used for preparing bisphenol A-free epoxy resins .

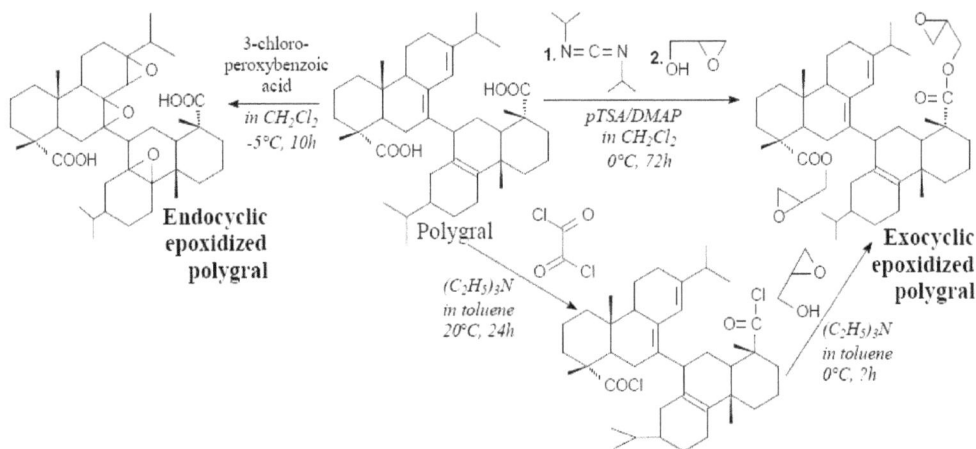

Scheme 7. Preparation of epoxidized rosin oligomers

Diglycidyl dehydroabietylamine is a yellowish sticky liquid. It can be synthesized from dehydro abietylamine, epichlorohydrin and sodium hydroxide according to Scheme 8, prior to filtration, washing, drying and vacuum distillation. It can be applied in epoxy resins exhibiting better thermal stability and higher glass transition temperatures than petroleum-based products.

Rosin maleimidodicarboxylic acid diglycidyl ether is a yellowish solid. It can be prepared from epichlorohydrin and dicarboxylic derivative of maleimide according to Scheme 9 prior to filtration, washing, drying and rotary evaporation. It shows higher glass transition temperature, modulus and thermal stability than its plant oil counterparts.

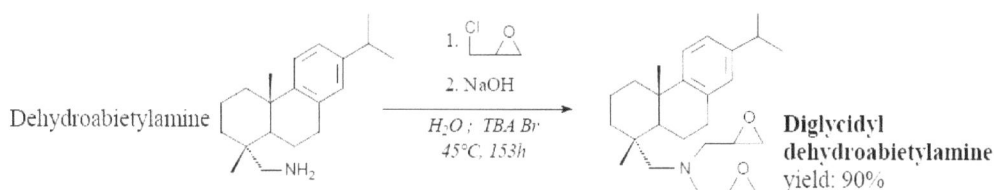

Scheme 8. Synthesis of diglycidyl dehydroabietylamine

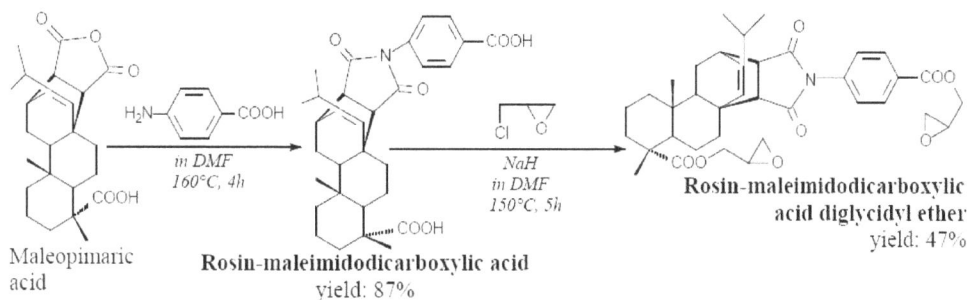

Scheme 9. Preparation of rosin maleimidodicarboxylic acid diglycidyl ether

Triester of maleopimaric acid and trimethylolpropane is a dark yellow solid. It can be prepared according to Scheme 10, prior to washing and vacuum distillation. It can be used in synthesis of rosin-based epoxy resin.

Epoxy resin having EEW = 199.68 g/eq based on maleopimaric acid and trimethylolpropane ester can be prepared according to Scheme 10, prior to extraction and vacuum distillation . It can be used directly as an epoxy compound, or transformed into acrylate resin.

Acrylate resin based on epoxidized maleopimaric-trimethylolpropane adduct can be prepared according to Scheme 10. It can be used as a resin crosslinked by styrene, methacrylated eugenol or methacrylated guaiacol.

Dehydroabietic ethyl methacrylate is a white powder. It can be prepared from dehydroabietyl chloride and hydroxyethyl methacrylate according to Scheme 11, prior to neutralization, drying, evaporation and chromatography. It can be widely used as a monomer in preparation of graft and block copolymers. It is worth noting, that a similar compound dehydroabietic ethyl acrylate is a yellow, viscous liquid, which can be prepared from hydroxyethyl acrylate and dehydroabietyl chloride prior to filtration, precipitation and vacuum distillation. It can be used for preparation of homo polymer with no declared application. Moreover, rosin ethyl acrylate can be used in preparation of bio-based graft copolymers of chitosan for controlled release applications.

Dehydroabietic hexyl acrylate is a solid. It can be synthesized from dehydroabietyl chloride, hexanediol and acryloyl chloride according to Scheme 12, and using separation methods such as filtration, precipitation, washing and vacuum drying. Its application is a soft acrylic monomer (glass transition temperature of -23°C), which can impart a flexibility to the integrated polymer.

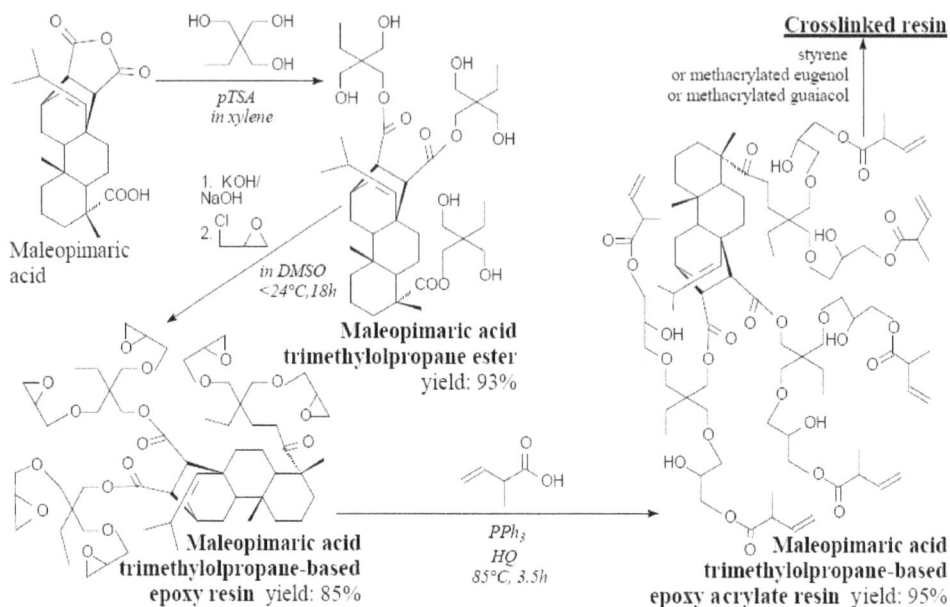

Scheme 10. Preparation of maleopimaric acid-trimethylolpropane ester-based epoxy acrylate resin crosslinked by unsaturated monomers.

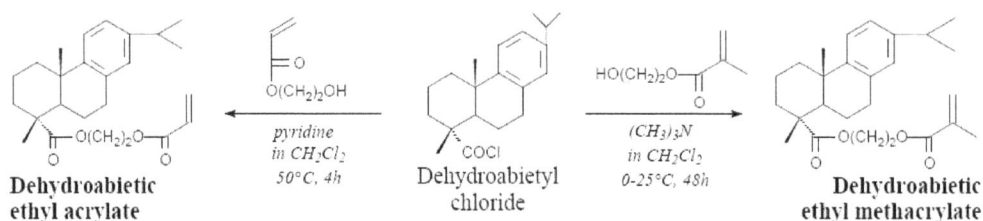

Scheme 11. Synthesis of dehydroabietic ethyl acrylates.

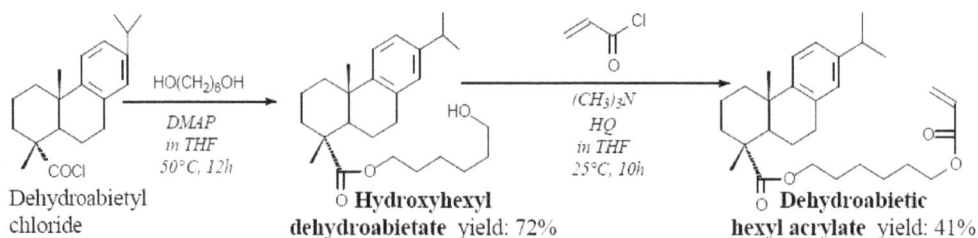

Scheme 12. Synthesis of dehydroabietic hexyl acrylate.

N-hydroxyethylacrylamide ester of dehydroabietic acid is a viscous, yellow liquid. It can be prepared from dehydroabietyl chloride and N-hydroxyethylacrylamide according to Scheme 13, prior to filtration, washing, drying and column chromatography. It can be used as a monomer in thermoset system with soybean oil-based resin for coating and adhesive applications

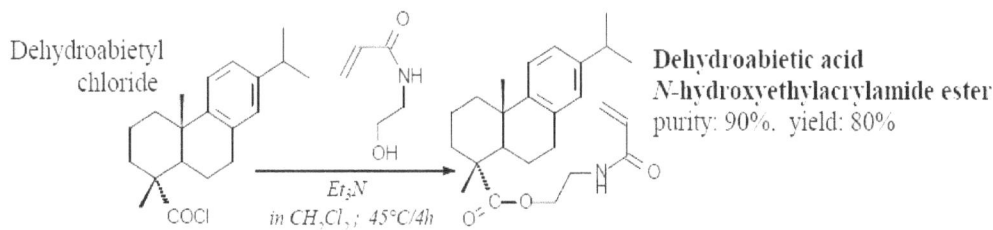

Scheme 13. Synthesis of N-hydroxyethylacrylamide ester of dehydroabietic acid

Rosin-based high adhesion polyurethane acrylate is a faint-yellow solid. It can be synthesized from hydrogenated rosin, isophorone diisocyanate and 2-hydroxyethyl acrylate according to Scheme 14, prior to precipitation, vacuum drying and column chromatography. Its application is an adhesive having a high polymerization rate, low volume shrinkage and high adhesion.

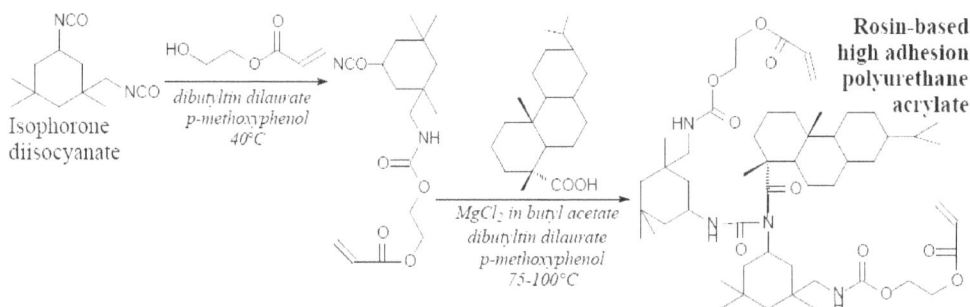

Scheme 14. Synthesis of rosin-based high adhesion polyurethane acrylate.

Rosin-based glycidyl methacrylate monomers are viscous liquids: brown rosin acid-glycidyl methacrylate and colorless dehydroabietylamine-glycidyl methacrylate. They can be prepared from glycidyl methacrylate and rosin or dehydroabietylamine according to Scheme 15, prior to the use of such separation techniques as washing, extraction and evaporation. They significantly improve thermal and mechanical properties of soybean oil-based thermosets. Moreover, they can be used in copolymerization with other acrylate monomers, or as an advanced tackifier in the UV-crosslinking pressure sensitive adhesives.

Ethylene glycol maleic rosinate (meth) acrylate can be prepared from maleated rosin, ethylene glycol and (meth) acrylic acid according to Scheme 16. It can be applied in styrene-acrylate copolymers increasing their thermal stability, as well as in preparation of molecular imprinted polymers for stationary phases used in high-performance liquid chromatography.

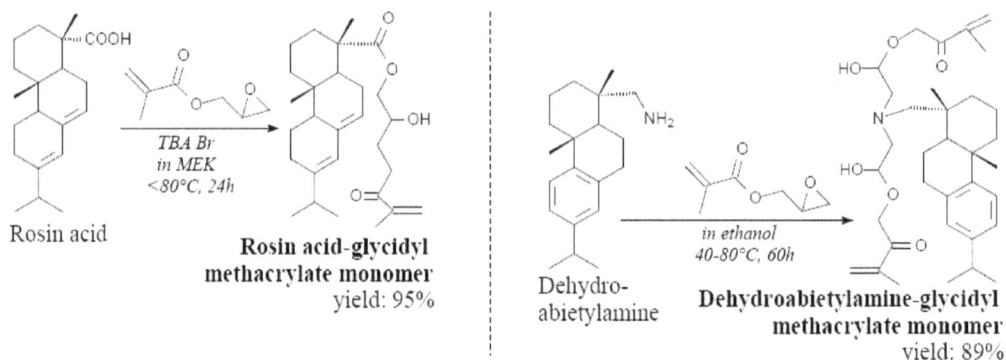

Scheme 15. Synthesis of rosin-based glycidyl methacrylate monomers

Diallyl acrylpimarate is a yellow liquid. It can be prepared from acrylpimaric acid and an allyl halide by different methods, according to Scheme 17, and using separation methods such as filtration, washing, extraction and evaporation. It can be applied as a monomer in polyester unsaturated resins from renewable resources, or in synthesis of aminated curing agent for epoxy.

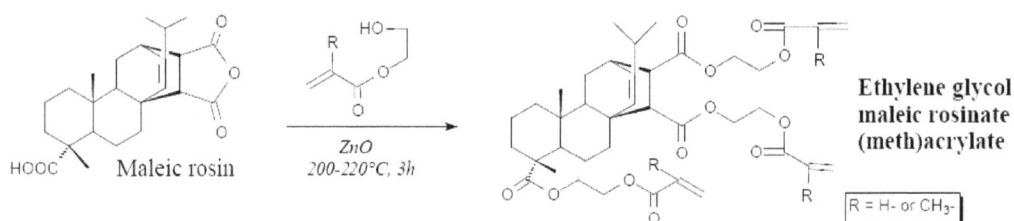

Scheme 16. Preparation of ethylene glycol maleic rosinate acrylate

Aminated diallyl acrylpimarate is a yellow-brown solid. It can be synthesized from diallyl acrylpimarate and cysteamine hydrochloride according to Scheme 17, prior to washing, extraction and evaporation. It can be applied as a resin or a curing agent for epoxy resins, giving them improved thermal and shape-memory properties.

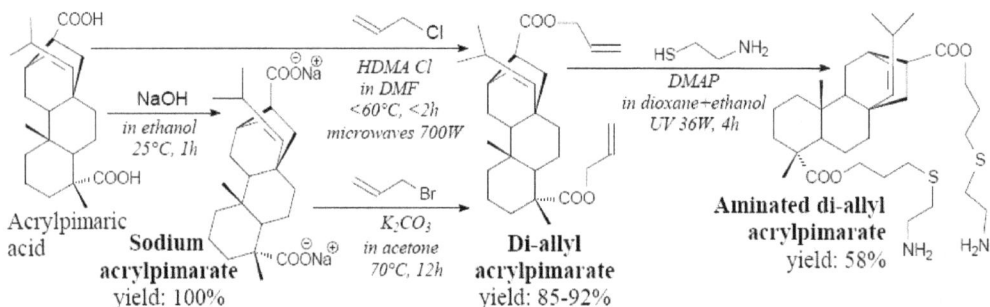

Scheme 17. Synthesis of diallyl acrylpimarate and its aminated derivative.

Sodium maleopimarate is a solid, which can be prepared from maleopimaric acid according to Scheme 18 prior to drying at 40°C. Its applications include the synthesis of triallyl maleopimarate.

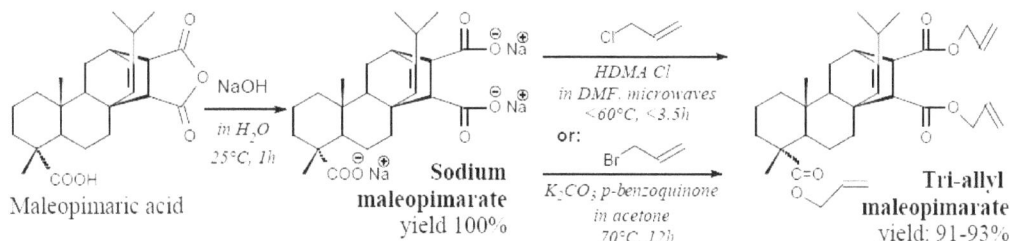

Scheme 18. Synthesis of triallyl maleopimarate.

In a similar way dehydroabietic acid salts can be synthesized for use as nucleating agents for isotactic polypropylene in a non-isothermal crystallization process, improving the crystallization temperature and accelerating the nucleation rate. Triallyl maleopimarate is a white viscous liquid, which can be prepared from sodium maleopimarate and allyl halide according to Scheme 18, prior to extraction, washing, filtration and vacuum distillation. It can be used as a monomer in UV-cured polymer films and coatings, giving them improved adhesion and mechanical properties, as well as in thermally cured fully bio-based resin systems exhibiting satisfactory thermal and mechanical properties.

Mono-allyl rosin derivatives have been also synthesized in recent years. Allyl rosinate can be prepared in aqueous or ethanol medium from rosin, sodium hydroxide and allyl chloride according to Scheme 19, prior to filtration and distillation. It can be potentially applied as an unsaturated monomer in copolymerization reactions as well as in UV-cured resins. On the other hand, allyl maleopimarate can be prepared in THF from maleopimaric acid, oxalyl chloride and allyl alcohol according to Scheme 19, and using such separation methods as vacuum distillation and column chromatography. It can be used in unsaturated resins and epoxy resins as a cross-linking agent.

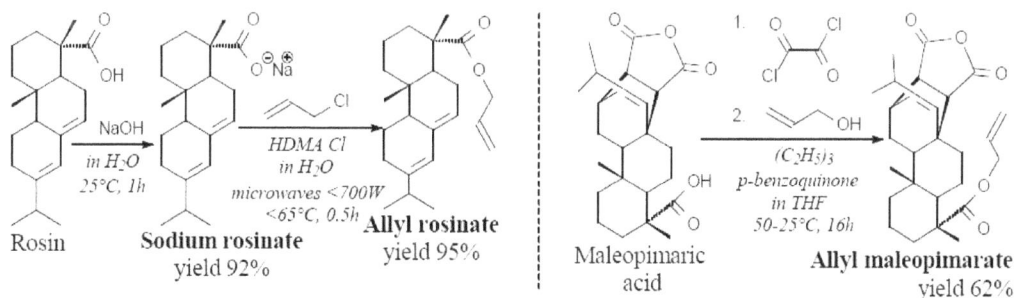

Scheme 19. Synthesis of mono-allyl rosin derivatives.

Benzoxazines are compounds consisting of a bicyclic group with an oxazine moiety. Maleopimaric acid imidophenol is a white, crystal solid. It can be synthesized from maleopimaric acid and 4-aminobenzoic phenol according to Scheme 20, prior to precipitation and drying. It can be used in synthesis of benzoxazine monomers.

Rosin-based benzoxazine monomers are orange solids. They can be synthesized from maleopimaric acid, paraformaldehyde and aniline or 4-aminobenzoic acid, according to Scheme 20, prior to filtration, washing and rotary evaporation. They can be polymerized into products of significant thermal stability.

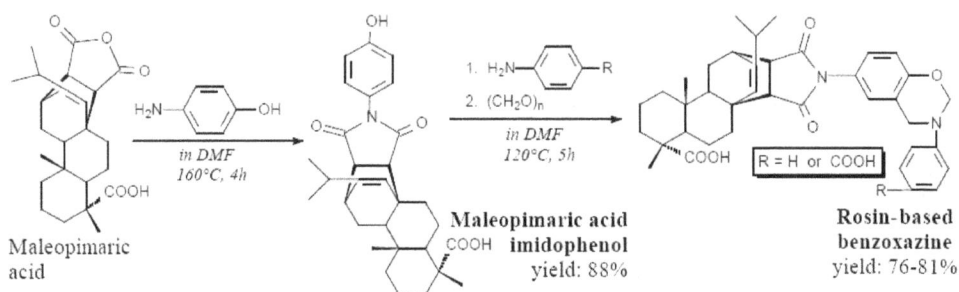

Scheme 20. Synthesis of maleopimaric acid imidophenol and its benzoxazines.

Dehydroabietylamine-guaiacol (brown powder, m.p. 104°C) and dehydroabietylamine-4- methylumbelliferone (yellow spherical crystal solid, m.p. 131°C) are fully bio-based benzoxazines. They can be synthesized from dehydroabietylamine via Mannich condensation, according to Scheme 21. They can compound resins with strong corrosion protection and thermal stability.

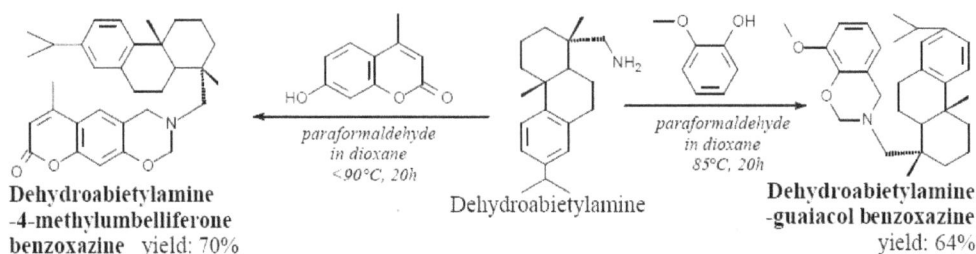

Scheme 21. Synthesis of fully bio-based benzoxazines

HARDENERS

Rosin-based hardeners, i.e. curing agents, are compounds containing anhydride, carboxyl or hydroxyl groups, which allow one to apply them in epoxy or urethane resin systems. In comparison with petroleum-based hardeners they bring improved endurance and thermal properties to resin systems. Moreover, rosin-based anhydrides are non-toxic, as opposed to conventional anhydride curing agents. Their preparation is simple and well described, and usually does not require the use of solvents. These strengths cause the great interest in use of rosin based chemicals as hardeners of conventional and bio-based resins resulting in several commercializations of eg., maleated rosins.

Methyl maleopimarate is a white solid. It can be prepared from abietic acid, iodomethane and maleic anhydride, according to Scheme 22, prior to recrystallization. It can be used as bio-based curing agent for epoxies, as well as in synthesis of binaphthyl-appended crown ethers derived from rosin.

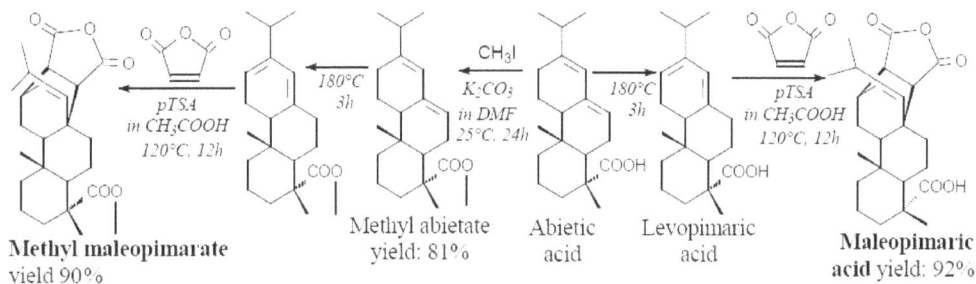

Scheme 22. Synthesis of maleopimaric acid and methyl maleopimarate

Rosin-maleimidopolycarboxylic acids are white/yellow or gray/brown powders. They can be prepared from maleopimaric acid with aspartic or 4-aminobenzoic acid, or rosin with 1, 1'- (methylenedi-4, 1-phenylene) bismaleimide, according to Scheme 23, before separation via such methods as precipitation, filtration, washing, drying and recrystallization. They have a potential to replace petroleum-based epoxy curing agents. Furthermore, rosinmaleimidodicarboxylic acid can be used for synthesis of rosin-based chain extender for polyurethanes, or epoxy resins.

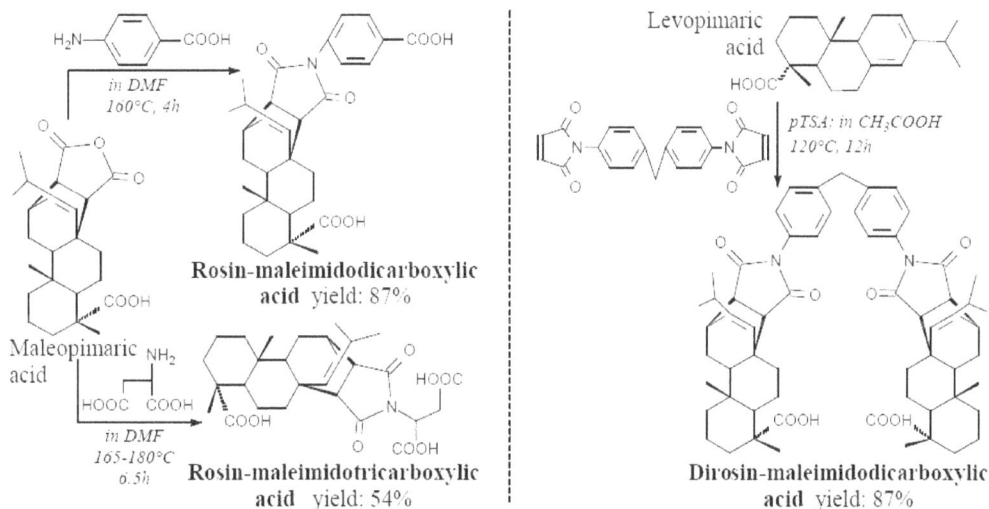

Scheme 23. Synthesis of rosin-maleimidodicarboxylic acids

Rosin-polycaprolactone flexible dianhydride are solids with a melting point depending on the length of oligoester diol chain. They can be synthesized from rosin, oligocaprolactone diols and maleic anhydride according to Scheme 24 and using vacuum evaporation as a separation method. Their application is bio-based curing agent for epoxy resins.

Tall-oil based polyol can be prepared from diethanolamine and tall oil containing up to 20 wt. % of rosin acids, according to Scheme 25, prior to vacuum evaporation of water. It can be used as chain extenders in ureaurethane elastomers, improving their thermal resistance and storage moduls.

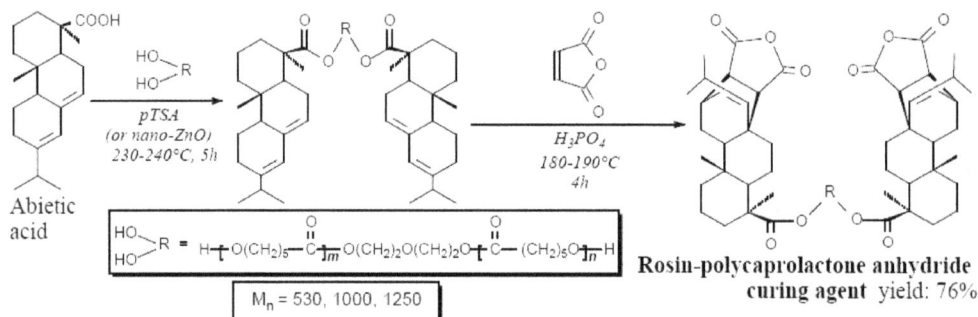

Scheme 24. Synthesis of rosin-polycaprolactone anhydride curing agent

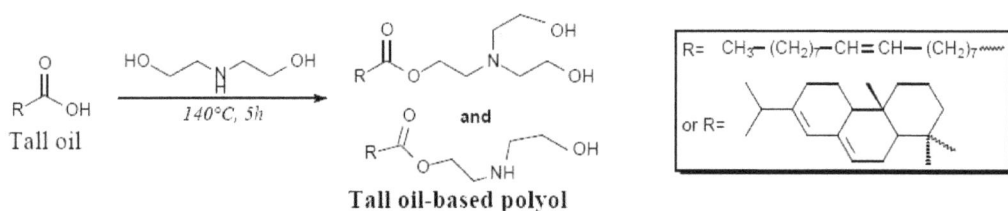

Tall oil-based polyol

Scheme 25. Synthesis of tall oil-based polyol

SURFACTANTS

Small/medium molecule rosin-based surfactants are ionic or non-ionic compounds that lower surface tension between different substances. In comparison with petrochemical counterparts, they are characterized by designable surface activity, affinity to many chemicals (especially cycloaliphatic and aromatic), non-toxicity, mild biocidal properties and enhanced thermal stability. Unfortunately, their syntheses are less well described than rosin-based resins and hardeners, eg. reaction yields are often unavailable. Moreover, the use of unsustainable chemicals in the mentioned reactions noticeably decreases the "green" aspect of these rosin derivatives. Therefore, it is an urgent need to undertake applied studies on these compounds, as well as to find more sustainable preparation processes in order to increase TRL of this group of rosin derivatives.

Dehydroabietyl phosphate diester is a yellow viscous liquid. It can be prepared from dehydroabietyl chloride, decanediol, and polyphosphorus acid according to

Scheme 26, prior to washing, vacuum evaporation and drying. It can be used as highly-active surfactant, phosphorus source and crystal growth control agent in synthesis of hydroxyapatite.

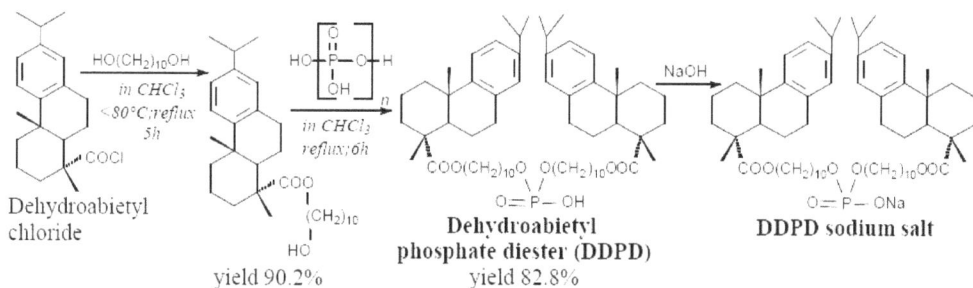

Scheme 26. Synthesis of dehydroabietyl phosphate diester sodium salt.

Rosin-based chloride can be prepared from maleated rosin and phosphorus trichloride according to Scheme 27, prior to evaporation. It can be used as a substrate for synthesis of rosinbased ester amines.

Rosin-based ester tertiary amine can be prepared from rosin chloride and N, N-dimethyl ethanolamine according to Scheme 27, prior to washing, drying, extraction, evaporation and recrystallization. Potential applications include drug carriers and surfactants. Its surface activity significantly increases in presence of rosin phosphate ester (Scheme 26).

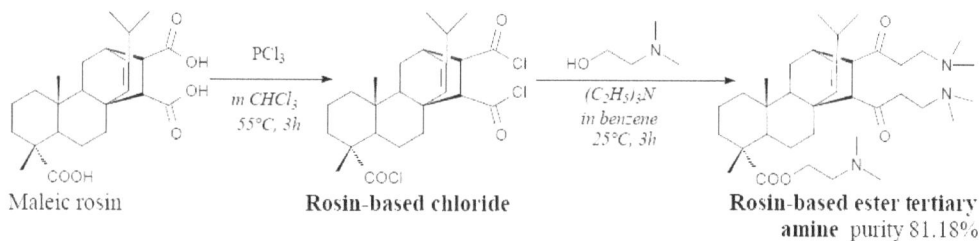

Scheme 27. Synthesis of rosin-based ester tertiary amine

Solid N-dodecyl-maleimidepimaric acid (C12-MPA) can be prepared from maleopimaric acid and dodecylamine according to Scheme 28, prior to washing, drying, filtration, rotary evaporation and purification via column chromatography. Sodium N-dodecylmaleimidepimaric carboxylate is a product of C12-MPA neutralization using NaOH, prior to evaporation, recrystallization and vacuum drying. It forms micelles of various shapes, depending on its concentration. It can be used in oil extraction, cosmetics and industrial washing.

Acrylpimaryl dichloride is an orange, sticky paste. It can be prepared from acrylpimaric acid according to Scheme 29, prior to solvent evaporation. It can be used in synthesis of surfactants, herbicides, fungicides and insecticides Acrylic rosin ester diethylamine tertiary amine surfactant is a sticky paste. It can be prepared from acrylpimaryl chloride, N, N-diethylethanolamine and hydrochloric acid according to Scheme 29. This product, in a mixture with soapnut saponin exhibits noteworthy surface activity and emulsification ability to apply in pharmacy, cosmetics and commodity chemicals.

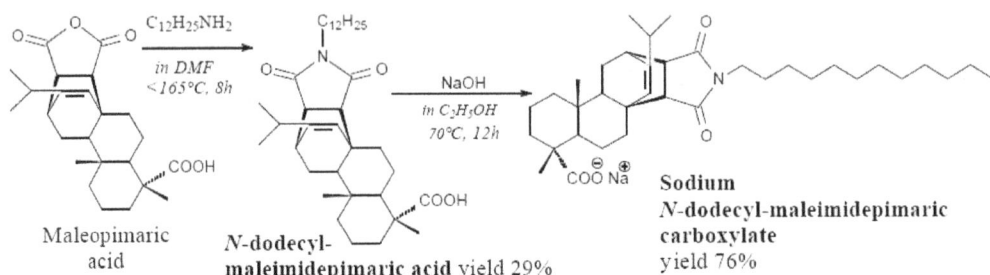

Scheme 28. Synthesis of sodium N-dodecyl-maleimidepimaric carboxylate

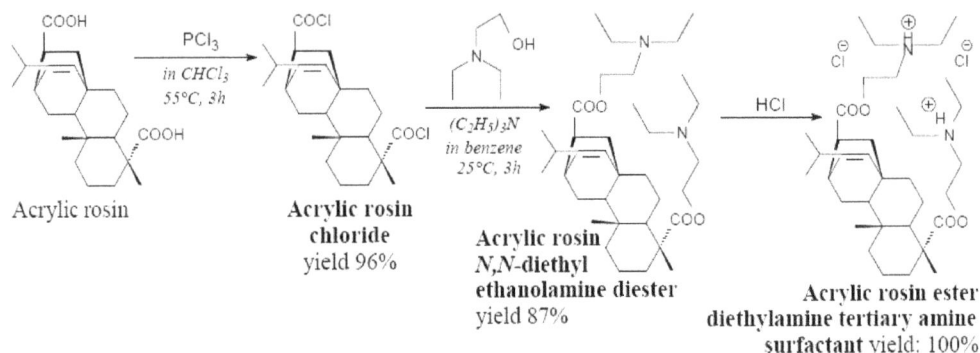

Scheme 29. Preparation of acrylic rosin ester diethylamine tertiary amine surfactant

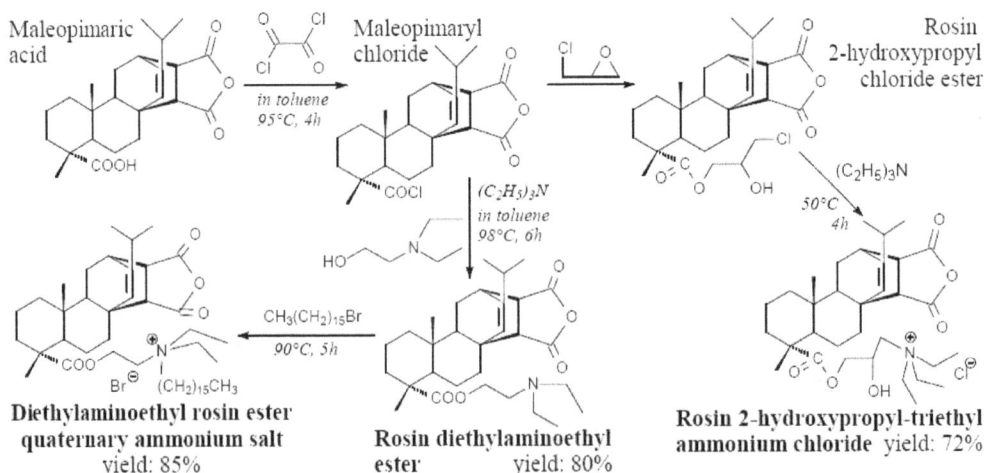

Scheme 30. Synthesis of quaternary ammonium salts of rosin esters

Quaternary ammonium salts of rosin esters are yellow solids. They can be prepared from maleopimaryl chloride and N, N-diethylethanolamine with hexadecyl bromide, or epichlorohydrin with triethylamine, according to Scheme 30, and using such separation methods as vacuum distillation, filtration, washing, drying, extraction and recrystallization.

Their potential applications include corrosion inhibitors dispersants for magnetite (Fe3O4) nanoparticles and inhibitors in protein aggregation processes. In addition, a similar rosinyl triquaternary ammonium salt having antifungal activity was also reported.

Another approach to introduce quaternary ammonium moieties into rosin is presented in Scheme 31. As it can be seen, rosin bisquaternary ammonium chloride can be prepared from rosin, ethanol, fumaric acid and epoxy quaternary ammonium salt, and using such separation methods as washing and drying. Thus obtained gemini surfactant has good surface activity and antifungal activity against fungi responsible for wood decay.

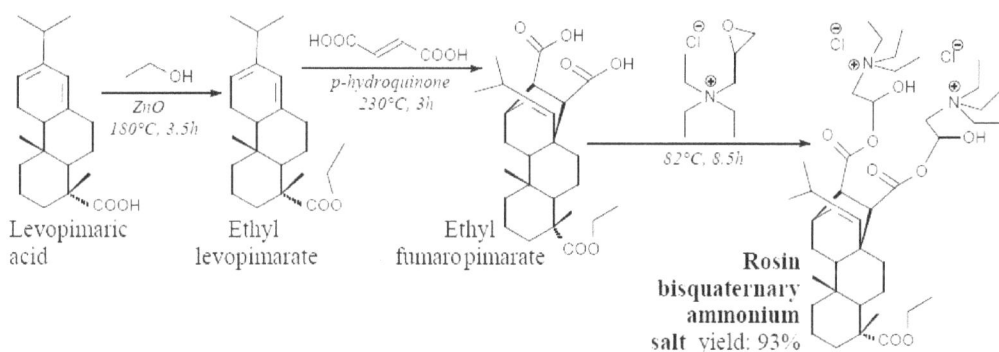

Scheme 31. Preparation of rosin bisquaternary ammonium salt

Dehydroabietyltrimethyl ammonium bromide is a yellow solid. It can be prepared from dehydroabietylamine, formic acid and methyl bromide according to Scheme 32, and using such separation techniques as washing, extraction and evaporation, drying, vacuum distillation and recrystallization. Such surfactant can be used for preparation of ordered porous titania, zirconia or silica with potential applications in catalysis and separation.

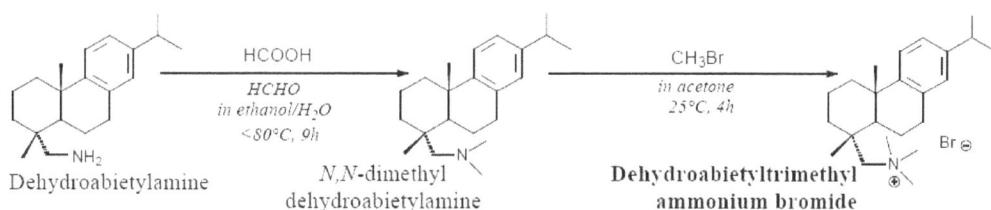

Scheme 32. Synthesis of dehydroabietyltrimethyl ammonium bromide

Another rosin-based gemini surfactants can be prepared from dehydroabietyl chloride, 3-(dimethylamino)-1-propylamine and α, ω-dibromoalkanes according to Scheme 33. The product separation methods include column chromatography, vacuum drying and recrystallization from ethanol/ethyl acetate. They can be used in preparation of three-dimensional mesoporous materials for separations, catalysis and drug delivery.

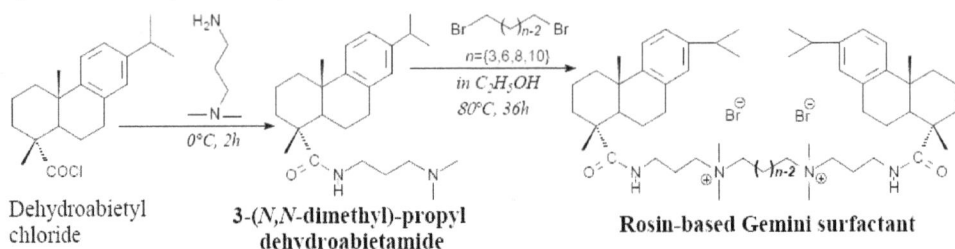

Scheme 33. Synthesis of rosin-based gemini surfactant.

Rosin-based cyclic tricarbonate is a brown solid. It can be prepared from maleopimaric acid triglycidyl ester and carbon dioxide according to Scheme 34, prior to washing and vacuum drying. Its applications include synthesis of quaternary ammonium salt derivatives and nonisocyanate polyurethanes. Rosin-based carbamate group-containing quaternary ammonium salt derivatives are brown solids. They can be prepared from rosin-based cyclic tricarbonate, N, Ndimethylaminopropylamine and alkyl bromides according to Scheme 34, and using vacuum drying and recrystallization. They exhibit strong antimicrobial properties.

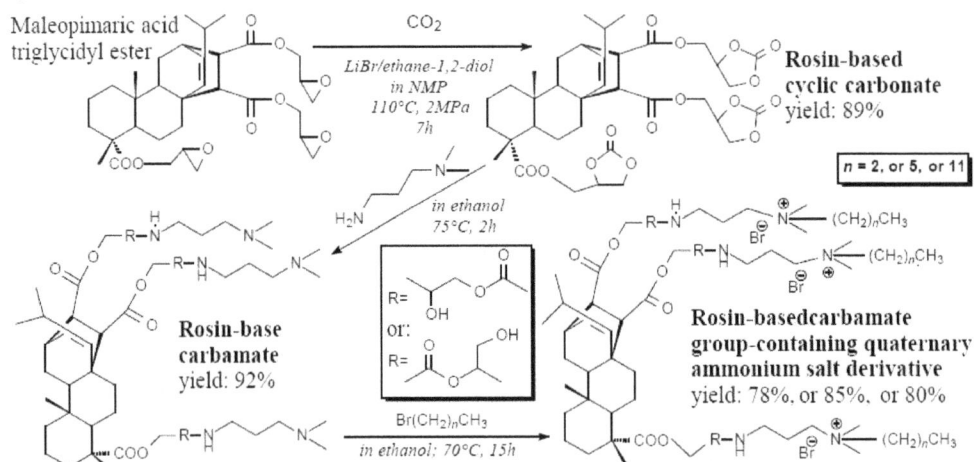

Scheme 34. Synthesis of rosin-based carbamate and carbamate group-containing quaternary ammonium salt derivatives.

Maleopimaric acid diethanolamide is a solid. It can be synthesized from maleopimaric acid and diethanolamine according to Scheme 35. It can be applied as a dispersant and a viscosity depressant in coal-water slurry.

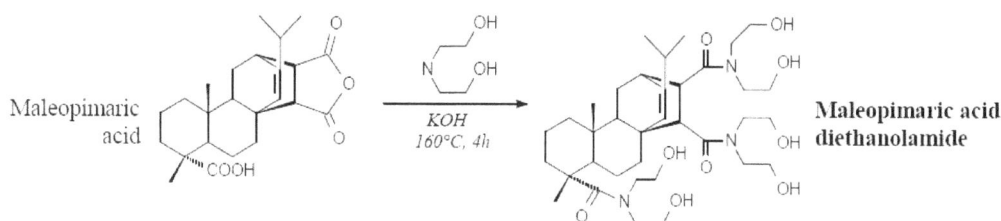

Scheme 35. Synthesis of maleopimaric acid diethanolamide

Acrylpiamric acid salts of calcium and zinc are solids. They can be prepared from acrylpimaric acid, sodium hydroxide and calcium chloride or zinc sulfate according to Scheme 36 and using such separation methods as washing, filtration and drying. They can be used as a stabilizer of poly(vinyl chloride) showing better thermal stability than commercial Ca/Zn stearate.

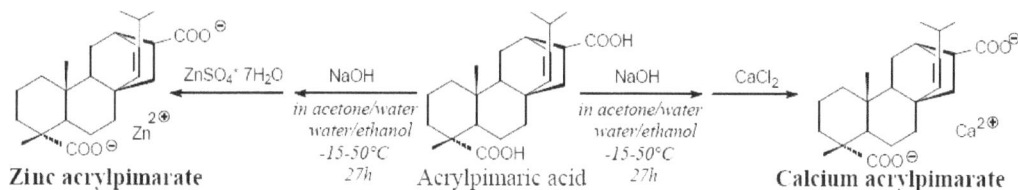

Scheme 36. Synthesis of acrylpimaric acid divalent metal salts.

Rosin-derived binapthyl-appended 22-crown-6 ether can be synthesized from methyl maleopimarate, sodium borate and sodium hydride according to Scheme 37. It can be used in highly enantioselective reactions because of its amines enantiomeric recognition ability. Very similar compounds can be also synthesized directly from maleopimaric or fumaropimaric acid.

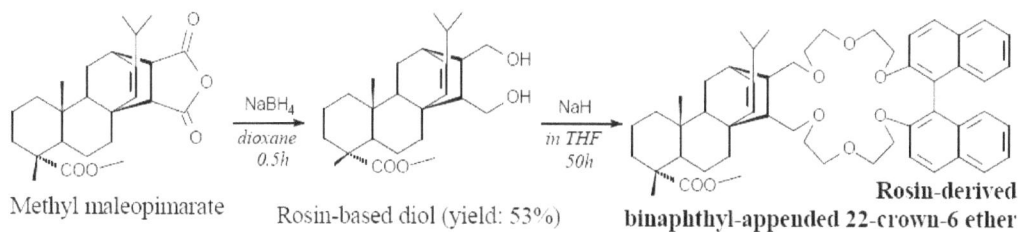

Scheme 37. Synthesis of rosin-derived binapthyl-appended 22-crown-6 ether

Biologically Active Compounds

Pure abietic acid isolated from rosin and its derivatives exhibit many potential activities ofinterest to the pharmaceutical industry, for example, antitumor, anti-inflammatory, antimycotic and anti-arteriosclerotic properties and uses in treating digestive canal, acute and   erosive gastritis, allergy, asthma, arthritis and psoriasis. In these subsection rosin-based chemicals with main applications as biologically active compounds can be found. They include both medications and biocides. Their synthetic routes are usually characterized in detail and, from the chemical point of view, their TRLs are rather high. It is worth noting, that although the sustainability of their syntheses is usually worse than for other rosin derivatives (due to the necessity of using more dangerous chemicals), it is still better than their petrochemical counterparts.

Methyl dehydroabietate is a white solid. It can be prepared from dehydroabietyl chloride and methanol according to Scheme 38, prior to vacuum evaporation and recrystallization. It can be used as a substrate in the synthesis of antimicrobial agents and medicines.

Methyl 7-oxodehydroabietate is yellow oil. It can be prepared from methyl dehydroabietate and chromium trioxide according to Scheme 38, followed by extraction, drying and column chromatography. It can be used as a substrate in synthesis of antimicrobial agents and medicines.

Dehydroabietic acid derivative QC4 can be prepared from methyl 7-oxodehydroabietate, phenylhydrazine hydrochloride, 1, 2-dibromoethane and N-methyl piperazine, according to Scheme 38. It shows antimicrobial properties and, moreover, induces gastric cancer cell death via oncosis and apoptosis. Another dehydroabietic acid derivative QC2 is reported to be able to inhibit skin cancer cell lines.

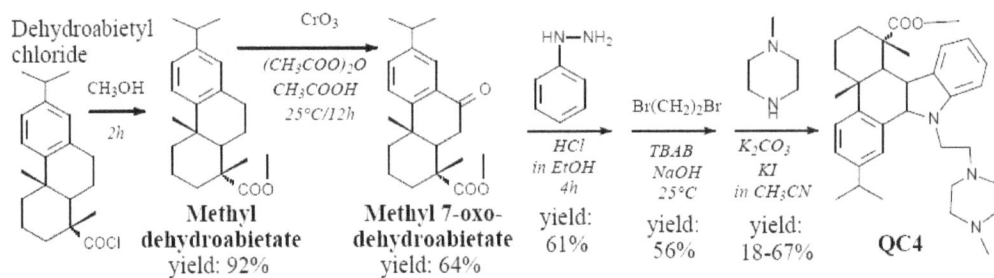

Scheme 38. Synthesis of dehydroabietic derivative QC4.

N-(2-methyl-naphthyl) maleopimaric acid diimides and their methyl esters are white solids (m.p. 215-290°C). They can be synthesized from maleopimaric acid, 2-methyl-1-naphtylamine and dimethyl sulfate according to Scheme 39, and using separation methods such as extraction, washing, drying, evaporation and recrystallization. They show significant antitumor cytotoxicity against several human cancer cell lines, especially NCI cells and MGC-803 cells.

Anticancer effects of various rosin derivatives, especially thioureas, were also investigated; they showed significantly cytotoxicity toward diverse human carcinoma cell lines.

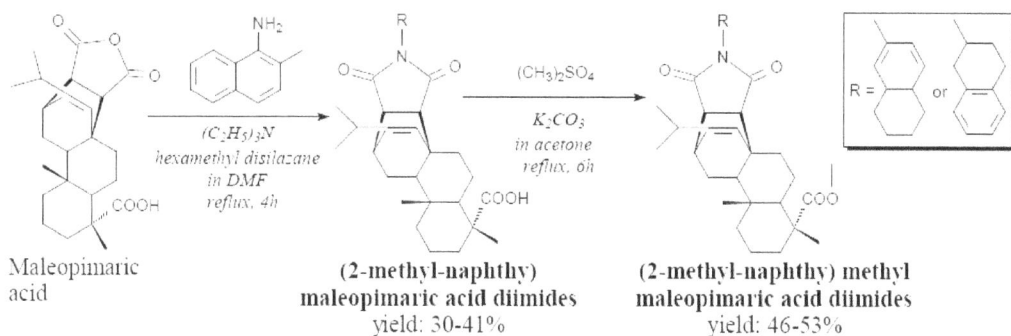

Scheme 39. Synthesis of N-(2-methyl-naphthyl) maleopimaric acid diimides.

Thiadiazole group-containing amides of dehydroabietic or acrylpimaric acids can be prepared from adequate rosin acid, thiosemicarbazide and an acyl chloride according to Scheme 40, and using such separation methods as filtration, vacuum drying, recrystallization and column chromatography. They can be potentially applied as insecticides.

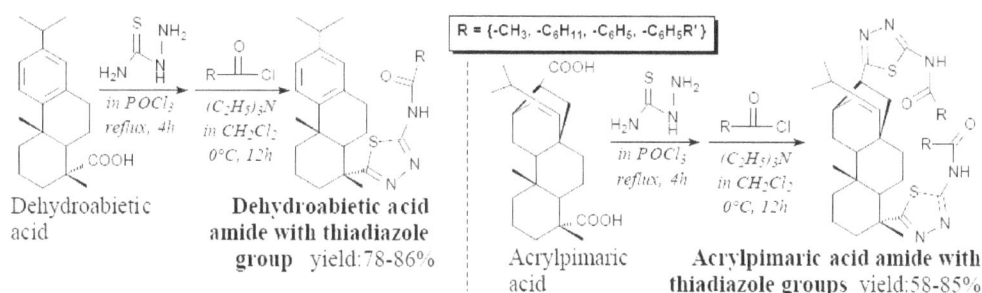

Scheme 40. Synthesis of rosin acid amides with thiadiazole groups.

Acrylpimaric acid-based aromatic diacylthioureas can be prepared from acrylpimaryl dichloride, potassium thiocyanate and aromatic amines according to Scheme 41, prior to vacuum evaporation and recrystallization. They may be applied as botanical herbicides showing higher activity than similar dicarboxamide, dihydrazone and diimide compounds.

Moreover, some acrylpimaryl diimides possess antibacterial activities against Gram-positive Staphylococcus aureus and Gram-negative Escherichia coli. Furthermore, some acrylpimaryl dicarboxamides show antibacterial properties against E. coli, whereas their activity against Gram-positive bacteria was significantly lower.

Solid acrylpimaryl quaternary ammonium salts can be synthesized from acrylpimaryl acid, oxalyl chloride, epichlorohydrin and volatile tertiary amines according to Scheme 41 and using such separation methods as vacuum distillation, extraction and recrystallization. They show fungicidal activity and can be applied in wood preservation. Very similar acrylpimaryl materials that can be applied as surfactants were also prepared. Another fungicidal rosin-based material can be prepared in similar way from rosin, epichlorohydrin and amines.

Maleopimaryl chloride is a yellowish solid. It can be prepared from maleopimaric acid and oxalyl (or thionyl) chloride, according to Schemes 30 and 42, prior to vacuum distillation and washing. It can be used as an intermediate in preparation of surfactants or fungicides

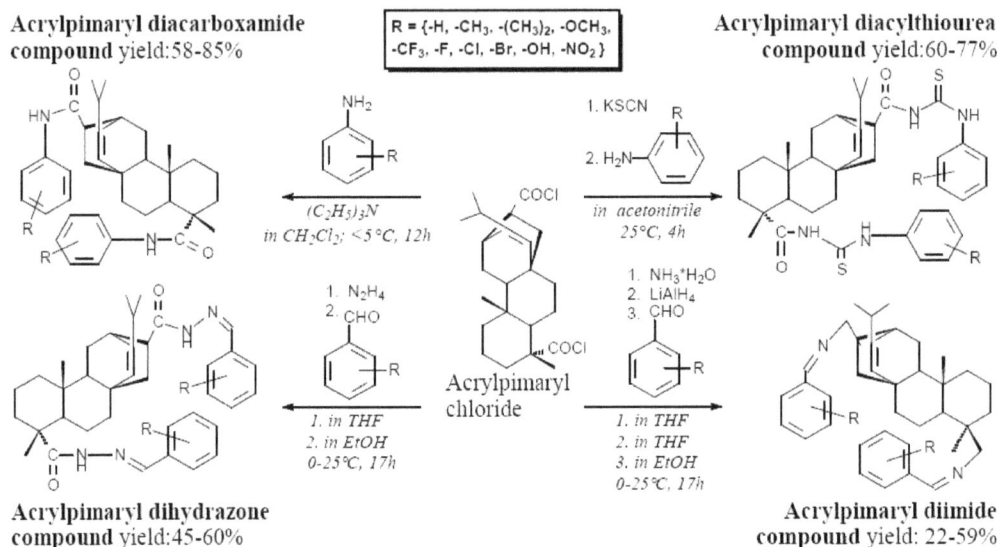

Scheme 41. Synthesis of acrylpimaric acid-based aromatic diacylthioureas, dicarboxamides, dihydrazones and diimides

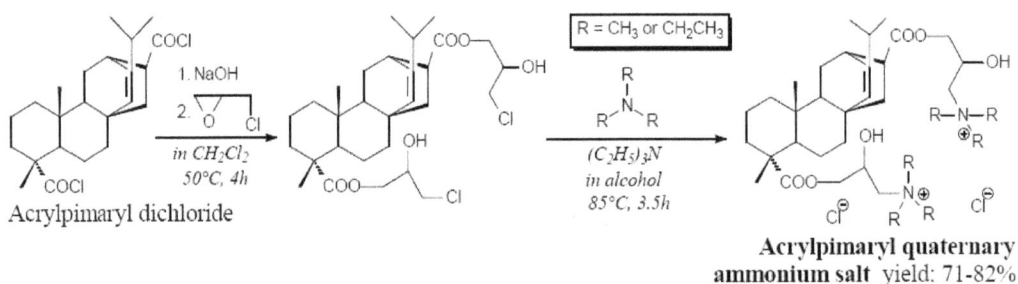

Scheme 42. Synthesis of acrylpimaryl quaternary ammonium salts.

Maleated rosin-based dithiourea compounds are yellow solids (m.p. 162–216°C). They can be prepared from maleated rosin acyl chloride, hydrazine and substituted benzoyl isothiocyanates, according to Scheme 43, and using separation methods such as rotary evaporation, recrystallization and column chromatography. They can be potentially used as fungicides.

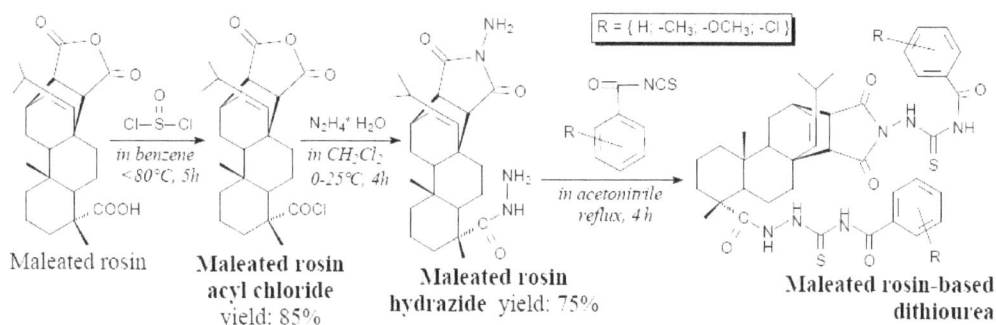

Scheme 43. Synthesis of maleated rosin-based dithiourea compounds

Glucose dehydroabietate can be prepared from dehydroabietic acid and glucose according to Scheme 44 prior to vacuum distillation, filtration and washing. Its potential application is as a surfactant in food industry.

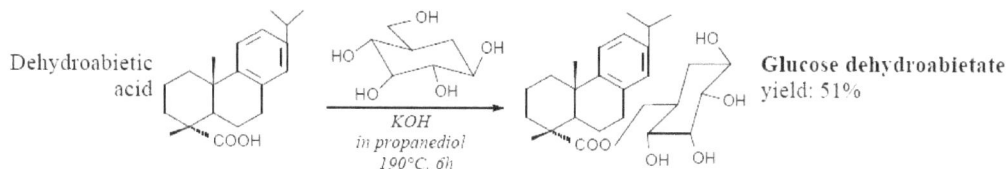

Scheme 44. Preparation of glucose dehydroabietate

Acrylic rosin cyclic diamide is a solid. It can be prepared from acrylpimaric acid and diethylene triamine according to Scheme 45 prior to washing and drying. It can be used as a fungicide for wood preservation.

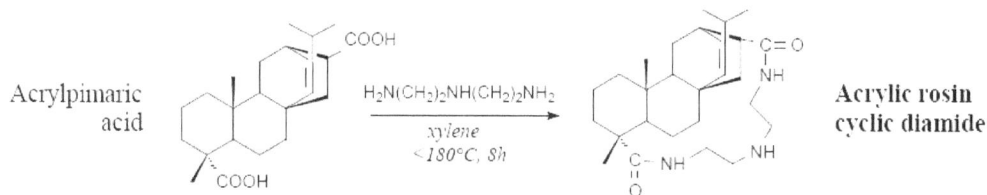

Scheme 45. Preparation of acrylic rosin cyclic diamide

OTHER SMALL/MEDIUM MOLECULE PRODUCTS

Hydroabietic acid can be prepared from rosin and hydrogen using palladium/SBA-15 mesoporous silica catalyst, according to Scheme 46. This catalyst is more efficient than others. Its applications include oxidation-resistant solvent-borne tackifiers and coatings, a substrate for esterification with glycerol, additives for polyvinylidene difluoride binders for lithium titanium oxide anodes. It can be used in high-performance liquid epoxy resins, for synthesis of high adhesion polyurethane acrylate. It is noteworthy, that non-catalytic decarboxylation of rosin can take place at temperatures above 200°C, and the main product of such decomposition is norabieta-8,11,13-triene.

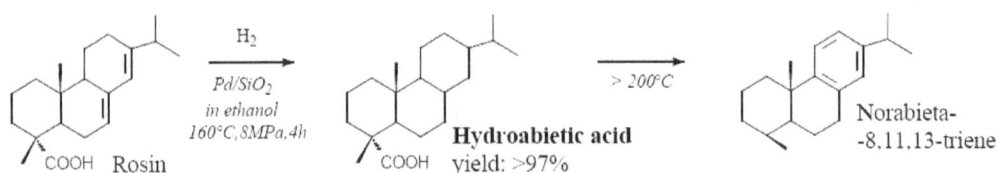

Scheme 46. Hydrogenation of rosin to hydroabietic acid.

Rosin-based standard-quality biodiesel can be prepared in a catalyst-free process from darkgrade rosin, heavy turpentine and supercritical methanol in supercritical CO2 as a green medium. Yields >93% were obtained after 3 h at 340°C, under a pressure of 11 MPa. Yield of >85% can be achieved using Pt/mesoporous aluminosilicate catalyst after 4 h at 300–350°C, 5 MPa. A yield >99% can be achieved using Ni/layered double hydroxide catalyst after <2 h at 190°C, 5 MPa. Catalytic cracking of rosin is also possible to carry out using acid-activated montmorillonite. Acrylpimaryl nitrile can be prepared from levopimaric acid and acrylonitrile according to Scheme 47, prior to extraction, filtration and precipitation. It can be used in synthesis of diacrylpimaryl ketone and acrylpimaryl amidoxime.

Acrylpimaryl nitrile amidoximes can be prepared from adequate acrylpimaryl nitriles and hydroxylamine according to Scheme 47, prior to precipitation, washing, and filtration and drying. They can be used to prepare bioactive thin films filled by magnetite nanoparticles for oil spill collecting and thorium ions removal.

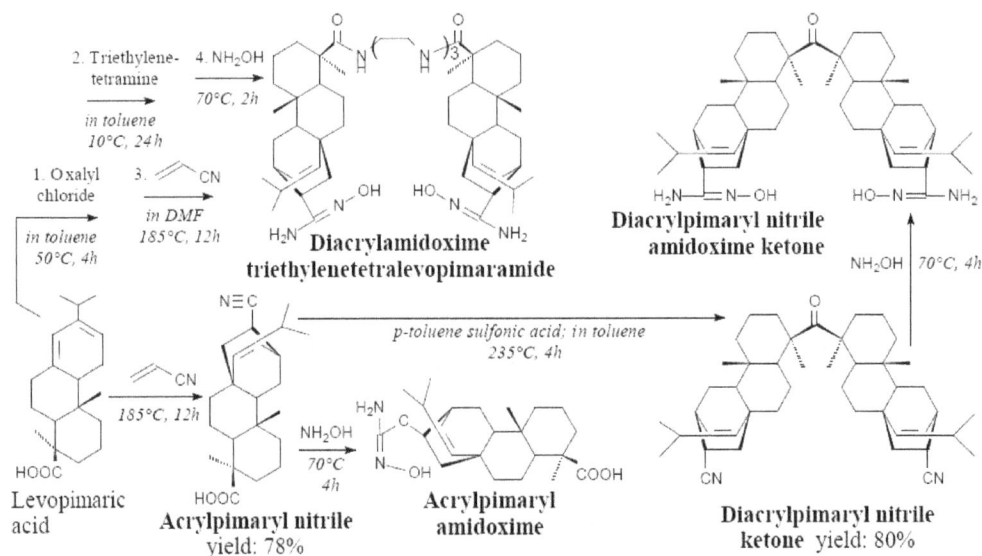

Scheme 47. Synthesis of acrylpimaryl nitrile and acrylpimaryl amidoximes

Rosin-oil dimer acids mixture can be prepared from rosin and industrial fatty oils according to Scheme 48, prior to washing and rotary evaporation. It can be used for preparation of a liquid thermal stabilizer. It is noteworthy, that dimerized rosin is usually produced separately and can be applied in acrylic adhesives to improve their wetting, adhesion and thermal stability.

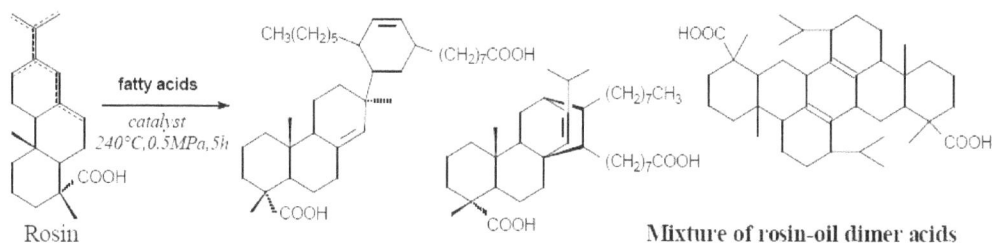

Scheme 48. Preparation of rosin-oil dimer acids mixture.

Solid rosin-based chain extender for polyurethanes can be synthesized from rosinmaleimidodicarboxylic acid, thionyl chloride and ethylene glycol, according to Scheme 49, and using such separation methods as vacuum distillation, washing and evaporation. Its application in shape memory polyurethanes improves shape recovery at >1000% strain up to 96%.

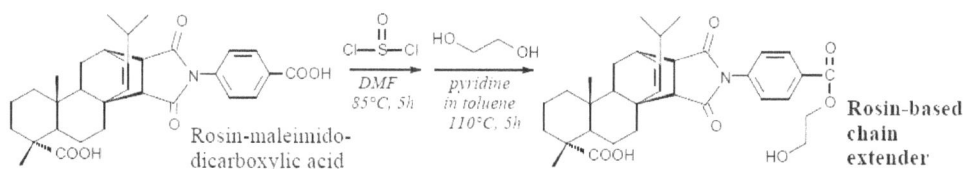

Scheme 49. Synthesis of rosin-based chain extender for polyurethanes

Chiral thioureas and thiouronium salts containing dehydroabietyl groups are well characterized white or yellow solids. They can be prepared from chiral amines (including dehydroabietylamine), carbon disulfide and butyl halides according to Scheme 50 and using such separation methods as filtration, washing, and vacuum drying, evaporation and column chromatography. They can be useful for the physical separation of racemic mixtures. Moreover, rosinderived thioureas can be used as enantioselective catalysts for many reactions. In recent years these were: Michael addition, tandem Michael/cyclization sequence, asymmetric Michael/ hemiketalization, asymmetric aza-Henry reaction, asymmetric tandem reaction, Mannich reaction, Friedel–Crafts alkylation, enantio- and diastereoselective asymmetric addition, as well as synthesis of chiral amines or N-protected β-amino malonates. In recent studies rosin-derived thioureas are dominant majority of all investigated rosin-derived catalysts. For now, reports on other rosin-based catalysts are rare.

Optically pure rosin-based chiral alcohols and their phosphorus derivatizing agents are white solids. They can be synthesized from maleopimaric acid in two main ways according to Scheme 51 and using such separation methods as washing, filtration, drying, evaporation, and recrystallization. They can be used in 31P-NMR-based determination of enantiomeric excess in solutions containing chiral alcohols and amines.

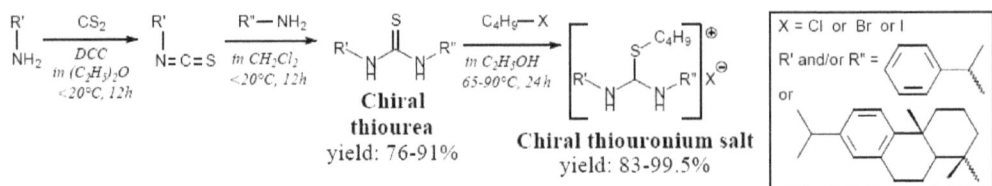

Scheme 50. Synthesis of chiral thioureas and their thiouronium salts.

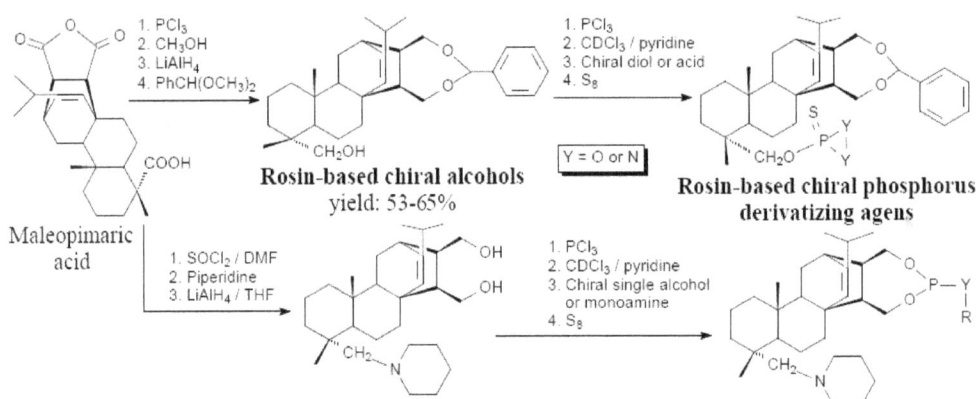

Scheme 51. Synthetic routes of rosin-based chiral alcohols and phosphorus derivatizing agents

Rosin-based molecular glass photoresists can be prepared from maleopimaric acid, hydroxylamine, 2-diazo-1-naphthoquinone-4-sulfonyl chloride and unsaturated compounds: vinyl ethyl ether, or dihydropyran, or cyclohexyl vinyl ether, according to Scheme 52 and using such separation methods as filtration, washing and vacuum drying. These materials can be applied in photolithography.

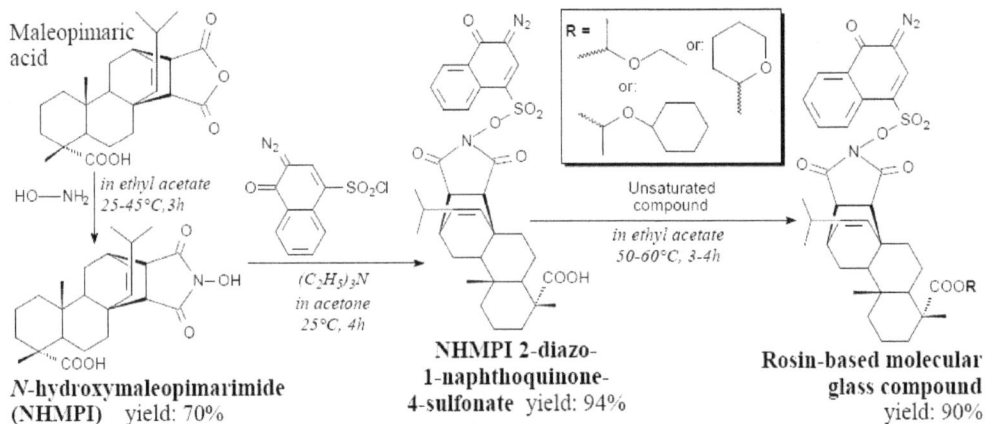

Scheme 52. Synthesis of rosin-based molecular glass compound

MACROMOLECULAR COMPOUNDS

This section describes rosin-derived macromolecular compounds with repeated units typical of macromolecules. It contains polymers, oligomers, macroinitiators and polymer functionalized materials. Nontoxicity, natural origin, low price, rigid structure, hydrophobicity, excellent thermal properties, anticorrosive performance, mild biocidal properties and stickiness are the most attractive reasons to use rosin derivatives in the preparation of polymer materials. A disadvantage of these processes in comparison with petrochemical counterparts is the relatively lower reactivity of rosin derivatives resulting from the steric hinderance of the diterpene skeleton and the usually lower purity of rosinbased intermediates. As a result, rosin-based polymer materials are usually characterized by distinct polydispersity and molecular weights rather far from their theoretical values. Therefore, their TRL is lower than for small/medium molecule compounds of rosin. It is noteworthy, that compared to non-renewable counterparts, the use of rosin significantly increases the sustainability of preparation processes according to Green Chemistry rules, so there is still high demand for basic and applied research in this field.

Polymers for Biomedical Applications

Poly(ethylene glycol) rosin esters are oligomers, which can be prepared from rosin, polyethylene glycol and maleic anhydride, according to Scheme 53 prior to drying at 40–70°C. Proposed applications include shells for controlled drug delivery and dental films for periodontitis treatment. Moreover, maleopimaric acid PEG esters can show carbon steel corrosion protection properties.

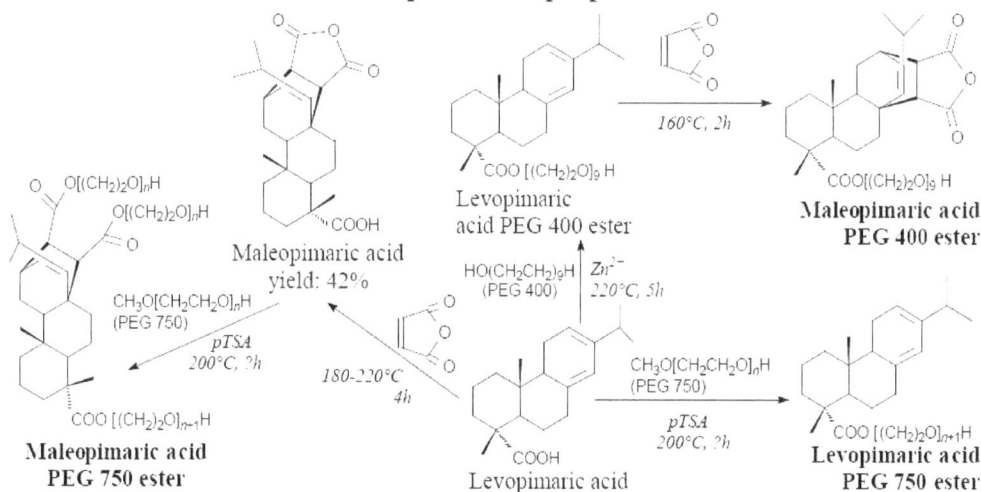

Scheme 53. Synthesis of poly(ethylene glycol) rosin esters.

Block copolymer of dehydroabietyl ethyl methacrylate and ethylene glycol with disulfide group can be prepared via ATRP according to Scheme 54, prior to neutralization, evaporation and precipitation. Potential applications include drug-delivery nanocarriers for cancer therapy.

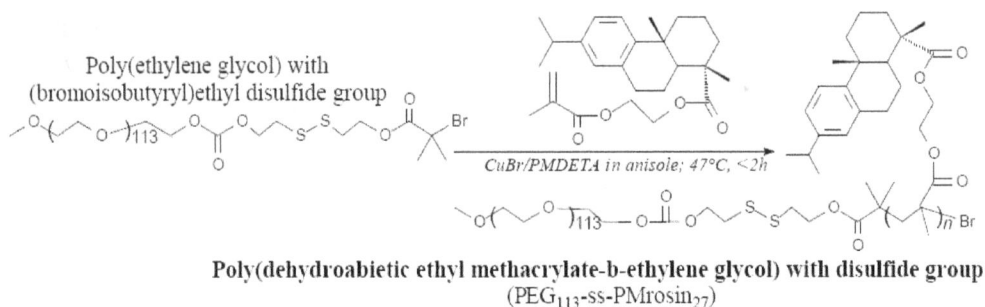

Scheme 54. Preparation of block copolymer of dehydroabietyl ethyl methacrylate and ethylene glycol with disulfide group

The rosin derivative quaternized poly-(N,N-dimethylaminoethyl methacrylate) can be prepared via "living" reversible addition-fragmentation chain-transfer polymerization (RAFT) from dehydroabietic acid, 3-chloropropanol, N,N-dimethylaminoethyl methacrylate and cumyl dithiobenzoate as a RAFT transfer agent, according to Scheme 55, and using gel chromatography, precipitation and vacuum drying as separation techniques. It can be used as amphipathic antibacterial agent in a wide variety of biomedical and general use applications.

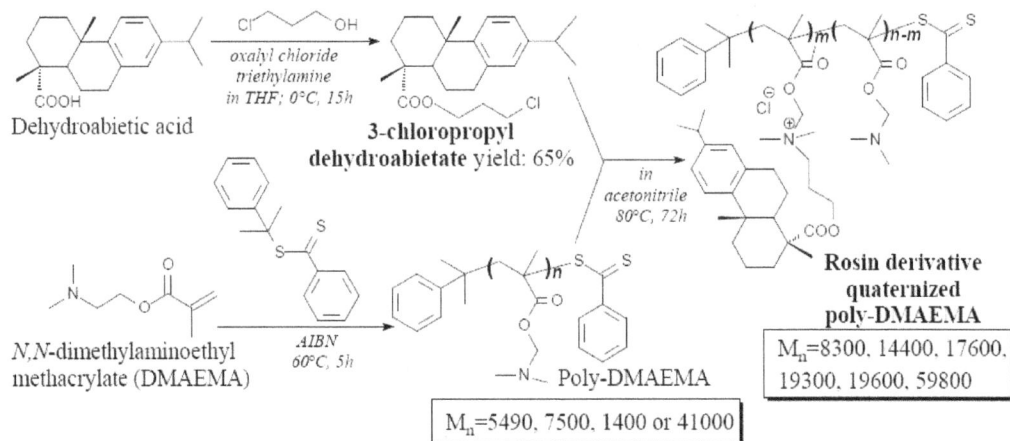

Scheme 55. Preparation of the rosin derivative quaternized poly (dimethylaminoethyl methacrylate)

ELASTOMERS

Poly(dehydroabietic ethyl methacrylate-β-n-butyl acrylate-β-dehydroabietic ethyl methacrylate) triblock copolymer can be prepared in ATRP polymerization from butyl acrylate, dehydroabietic hydroxyethyl acrylate and diethyl meso-2,5-dibromoadipate according to Scheme 56, prior to diluting, neutralization, evaporation, precipitation and vacuum drying. Its application is sustainable thermoplastic elastomers.

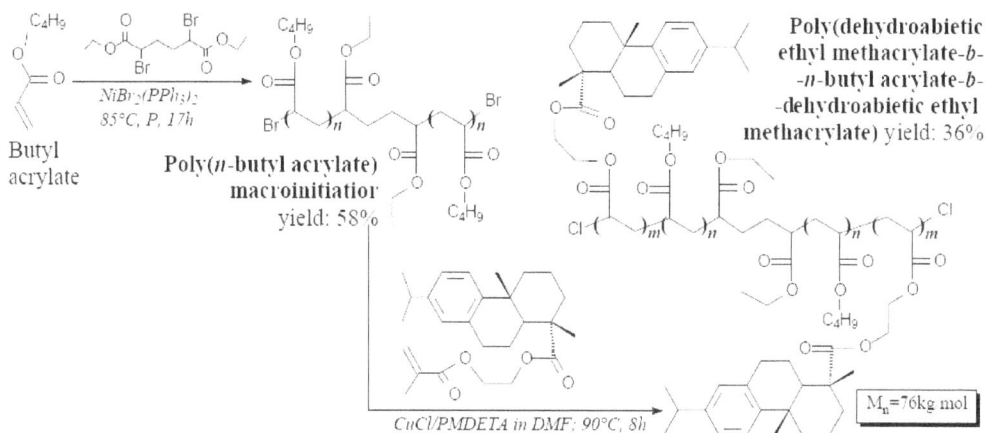

Scheme 56. Synthesis of rosin-based triblock copolymer

Cellulose/rosin ATRP macroinitiators can be prepared from dehydroabietic acid, cellulose and 2-bromoisobutyryl bromide, according to Scheme 57, prior to drying at 40°C. It is used for the preparation of graft copolymers.

Rosin-acid-modified ethyl cellulose/butyl acrylate graft copolymer can be prepared in ATRP reaction from dehydroabietic acid, cellulose, 2-bromoisobutyryl bromide and butyl acrylate according to Scheme 57, prior to sorption of CuBrx and precipitation in methanol. Potential applications include thermoplastic elastomers and coatings with UV absorption property.

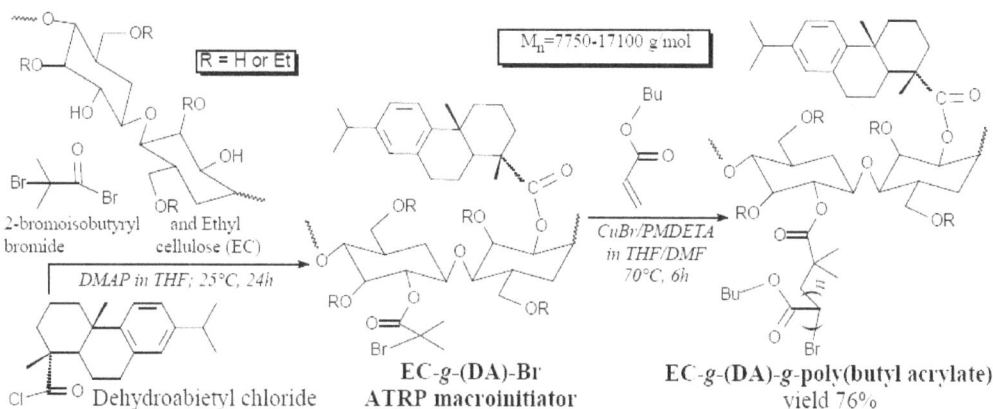

Scheme 57. Synthesis of graft copolymer of rosin-modified ethyl cellulose and butyl acrylate

Cellulose grafted by copolymer of rosin acid ethyl methacrylate and alkyl (meth) acrylate can be prepared via ATRP, according to Scheme 58, prior to sorption and precipitation in methanol. Potential applications include "green" thermoplastic elastomers having significant hydrophobic, thermal and mechanical features.

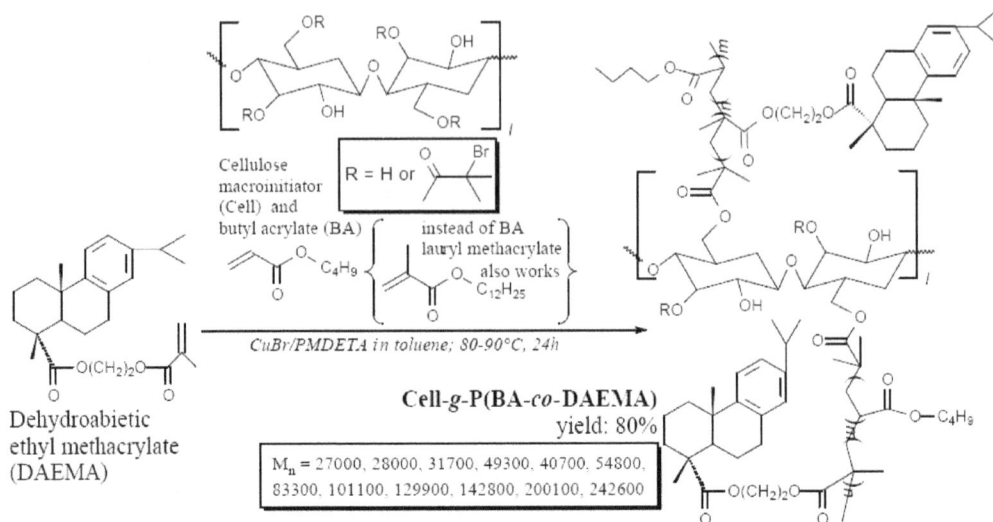

Scheme 58. Synthesis of graft copolymer of cellulose, rosin acid and butyl acrylate (or lauryl methacrylate)

Rosin alcohols are colorless solids, which can be prepared in several ways according to Scheme 59, and using such separation methods like evaporation, extraction, washing and drying. They can be used in preparation of cross-linked structures with acrylamide and N, N'- methylenebisacrylamide, as well as norbornene-based monomers.

Rosin-norbornene monomers, i.e. dehydroabietanyl norborn-5-ene-2-carboxylate and 4- ((norborn-5-ene-2-carbonyl)oxy)butyl dehydroabietate, are viscous oily liquids. They can be synthesized from dehydroabietic acid derivatives and norbornenecarboxylic acid according to Scheme 59, prior to evaporation, washing, drying and column chromatography. Its application is polymerization, or block copolymerization with norbornene, via "living" ring-opening metathesis polymerization (ROMP) process.

Homopolymers of rosin-modified norbornene can be synthesized via ROMP process according to Scheme 59. Moreover, triblock and pentablock copolymers with norbornene segments can be prepared. They can be applied as bio-based thermoplastic elastomers showing well-designed architecture and high elastic recovery.

Rosin-based waterborne polyurethanes can be prepared from maleopimaric acid, diethylene glycol, polyether glycol, toluene diisocyanate, dimethylol propionic acid and trimethylamine according to Scheme 60, and using such separation methods as vacuum drying and rotary evaporation. Such polymers can be also synthesized using fumaropimaric rosin instead of maleopimaric acid. These materials exhibit excellent mechanical properties, thermal stability, water resistance, antimicrobial properties against Gram-negative Escherichia coli and Gram-positive Staphylococcus aureus and an affinity for cellulose nanocrystals, which allows applying them in various biomass-based polymer and composite materials.

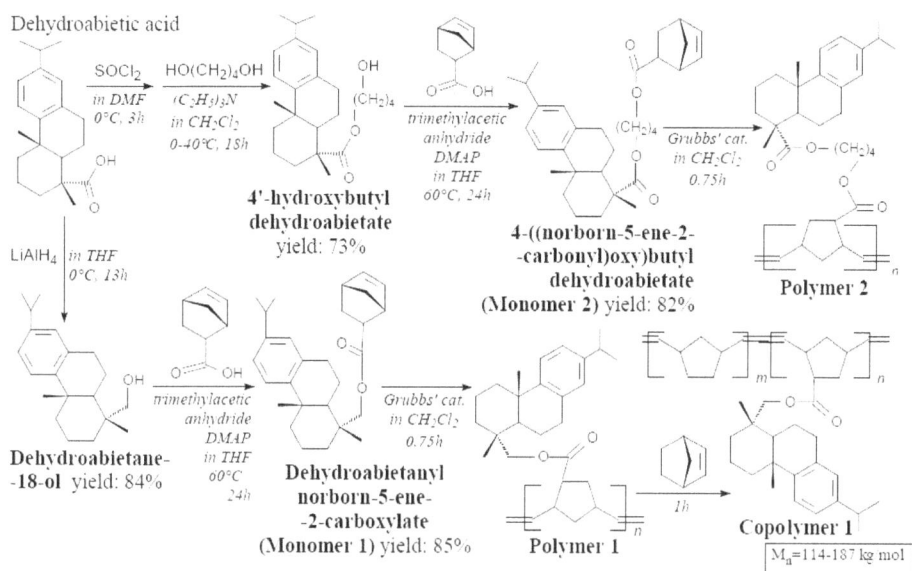

Scheme 59. Synthesis of rosin-norbornene monomers and polymerization of them

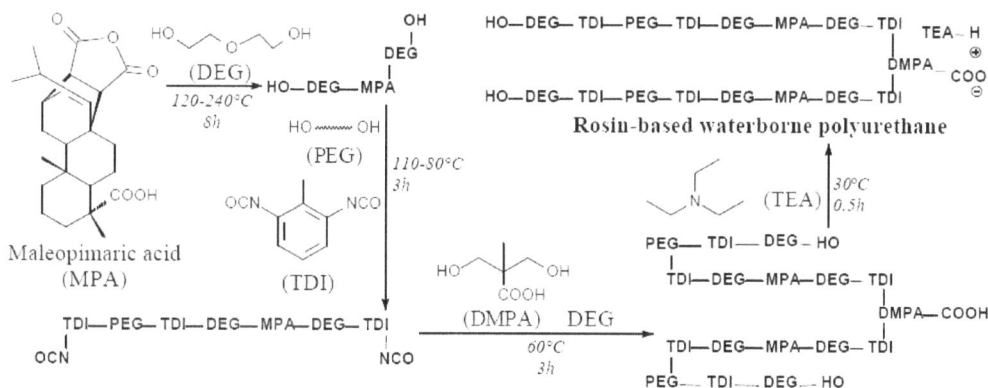

Scheme 60. Preparation of rosin-based waterborne polyurethane

COATINGS AND ADHESIVES

Rosin-modified poly(acrylic acid) is a solid. It can be prepared from poly(acrylic acid) and abietic acid according to Scheme 61, prior to washing and vacuum drying. It can be applied as an excellent binder for silicon-graphite negative electrodes in lithium-ion batteries.

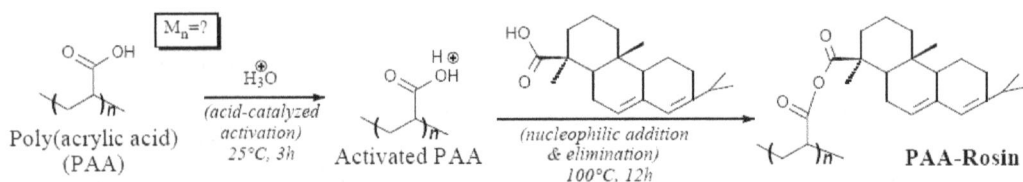

Scheme 61. Preparation of rosin-modified poly(acrylic acid).

N-dehydroabietic acrylamide is a white solid. It can be synthesized from dehydroabietylamine and acryloyl chloride according to Scheme 62, prior to washing and vacuum distillation. It can be used as a bio-based acrylic monomer in compolymerization processes instead of rigid petroleum based monomers.

Rosin and soybean oil-based acrylic copolymers can be prepared from N-dehydroabietic acrylamide and acrylated epoxidized soybean oil according to Scheme 62. They are characterized by considerable hydrophobicity and heat resistivity and also other properties comparable with similar petroleum-based materials.

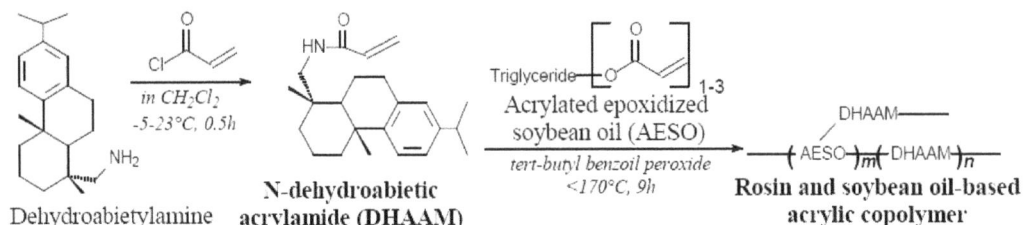

Scheme 62. Preparation of rosin and plant oil-based acrylic copolymers

Rosin and POSS-based non-isocyanate polyurethanes can be prepared from rosin-based cyclic carbonate, polyamine and cyclic carbonate-functionalized POSS according to Scheme 63. Their applications include coatings showing improved water tolerance, hardness and thermal stability.

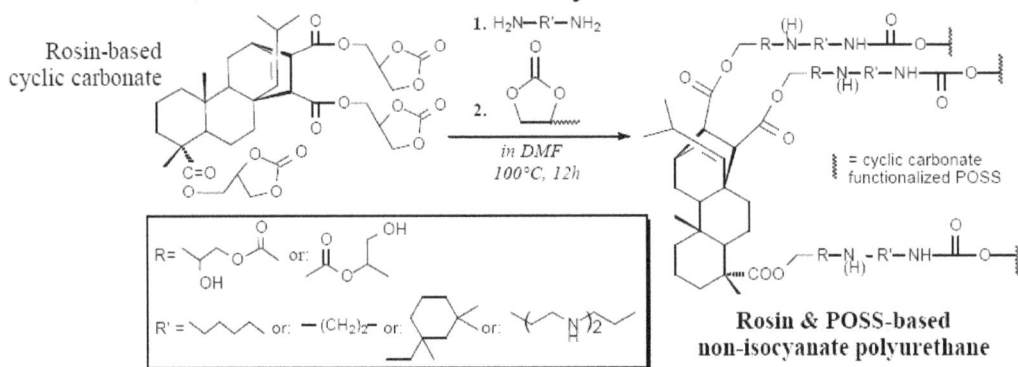

Scheme 63. Preparation of rosin & POSS-based non-isocyanate polyurethane

Rosin can be introduced as a chain extender into polyurethanes to obtain rosin-based urethaneamide hard segments, according to Scheme 64. A potential application of the prepared materials is as sealants for non-invasive disc regeneration surgery. Physical mixtures of rosin and 1, 4- butanediol were also investigated for this use.

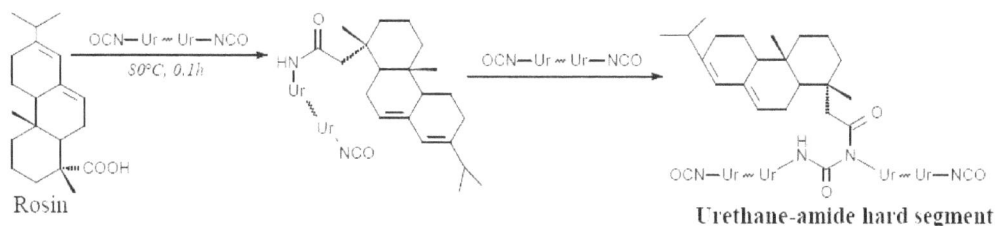

Scheme 64. Preparation of rosin-modified urethane-amide hard segments

Maleopimaric acid-modified polyester polyol aqueous dispersion can be prepared from maleopimaric acid, adipic acid, isophtalic acid, 5-sulfoisophtalic acid, neopentyl glycol and trimethylolpropane, according to Scheme 65, prior to dissolving/dispersing in water and diethylene glycol monoethyl ether acetate as a cosolvent. It can be applied in two-component waterborne polyurethane materials and coatings showing improved thermal stability, hardness and resistances to ethanol and water.

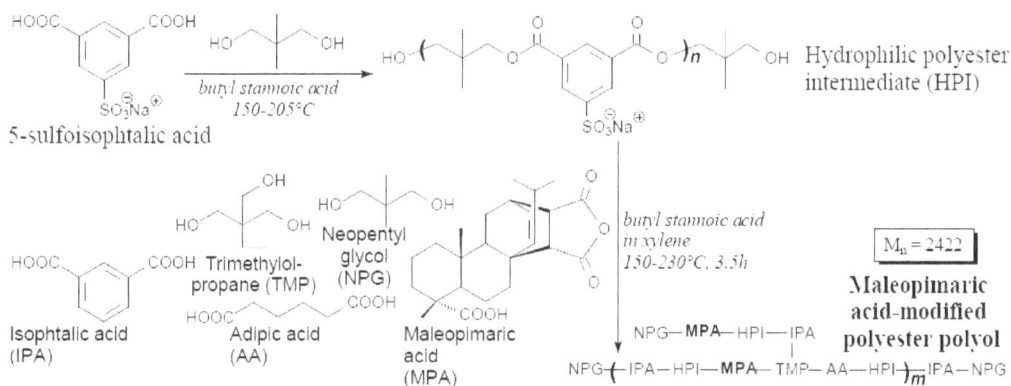

Scheme 65. Preparation of hydrophilic maleopimaric acid-modified poliester polyol

SURFACTANTS

Rosin-based comb-like polymeric surfactants can be prepared from rosin glycidyl methacrylate and methacrylate polyethylene glycol ester according to Scheme 66, prior to vacuum drying, precipitation, dialysis and freezing. Their application includes preparation of pymetrozine water suspension concentrates.

Polyesters of acrylated rosin and polyethylene glycols can be prepared according to Scheme 67. They can be used as surfactants in preparation of stable emulsions.

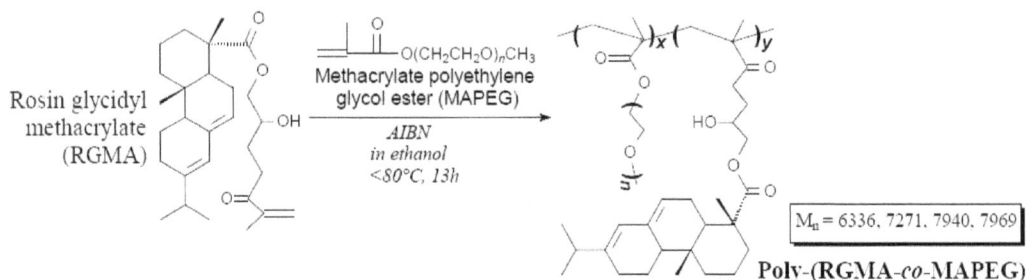

Scheme 66. Preparation of comb-like surfactants

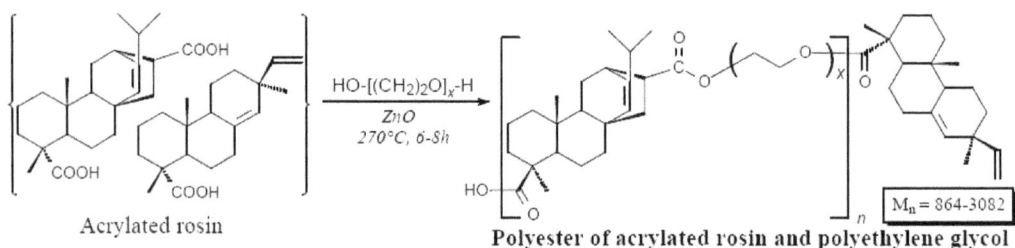

Scheme 67. Preparation of acrylated rosin/polyethylene glycol polyester

Rosin imide polyethers are light brown solids. They can be prepared from rosin, poly(ethylene glycol) and polyamines according to Scheme 68, and using such separation methods as washing, drying, precipitation and filtration. They can be applied as petroleum crude oil sludge dispersants.

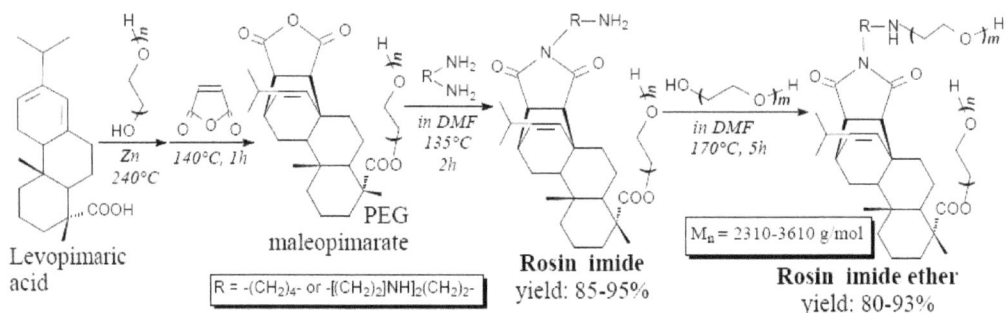

Scheme 68. Preparation of non-ionic rosin imide polyethers

Sorbents

Rosin-poly(acrylamide) star copolymers can be prepared according to Scheme 69, prior to Soxhlet extraction using acetone and drying. They can be used for wastewater treatment and as a matrix for Fe3O4 nanoparticles.

Linear rosin-modified cationic poly(acrylamide) can be prepared from dehydroabietyl chloride, bromopropan-1-ol, methyldiallylamine, acrylamide and diallyl dimethyl ammonium chloride in a few-step process, according to Scheme

70, followed by operations such as filtration, washing, drying and recrystallization. Its utilization may be in flocculation processes.

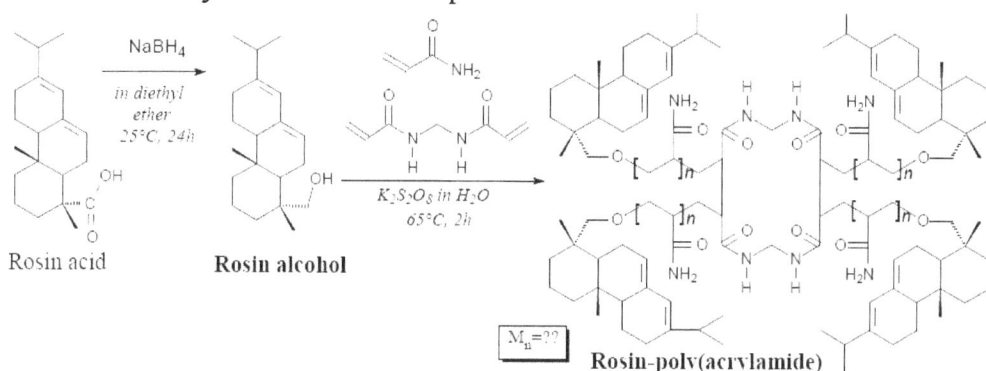

Scheme 69. Synthesis of rosin alcohol and rosin-poly(acrylamide) star copolymer

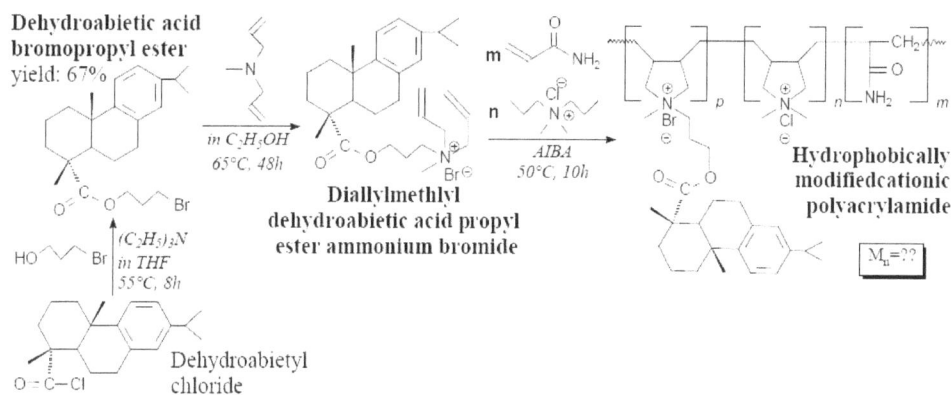

Scheme 70. Preparation of linear rosin-poly(acrylamide) copolymer.

Rosin tetra ethylenepentamine amide is a solid. It can be prepared from maleopimaric acid and tetraethylenepentamine according to Scheme 71, prior to precipitation, washing and lyophilization. It creates nano-micelles in aqueous solutions and can act as a sorbent and sinking agent in metal ions removal.

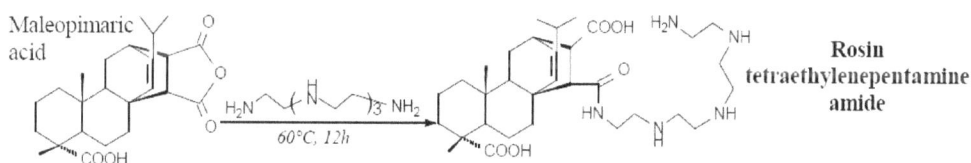

Scheme 71. Preparation of rosin-tetraethylenepentamine amide

Organosilicons

Rosin glycidyl ester is a brown, viscous liquid. It can be synthesized from rosin and epichlorohydrin according to Scheme 72 without further purification. It can be used for preparation of cross-linking agent for silicone rubber.

Rosin-modified room-temperature-vulcanized silicone rubber can be prepared from rosin glycidyl ester, amino propyltrietoxysilane, tetra etoxysilane and hydroxyl-terminated polydimetoxysilane according to Scheme 72. Rosin-modified silicones show significantly better thermal and mechanical properties in comparison with unmodified silicone.

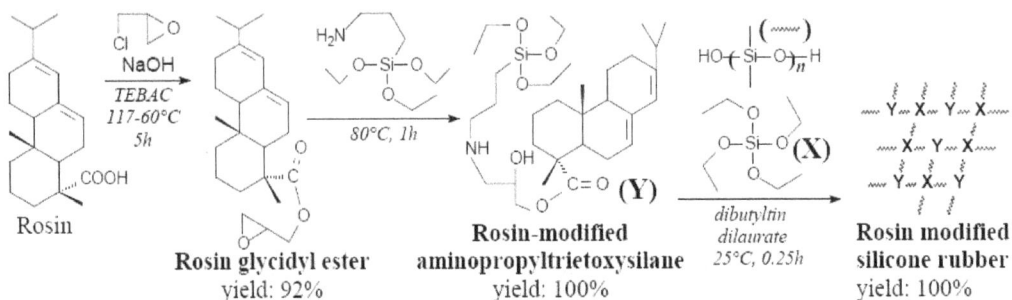

Scheme 72. Preparation of rosin-modified room-temperature-vulcanized silicone rubber

Maleated rosin-modified vinyl fluorosilicone resin can be prepared from maleopimaric acid and siloxanes according to Scheme 73 prior to vacuum evaporation. It can be used in preparation of fluorosilicone rubber, having improved mechanical and thermal properties in comparison with unmodified sample.

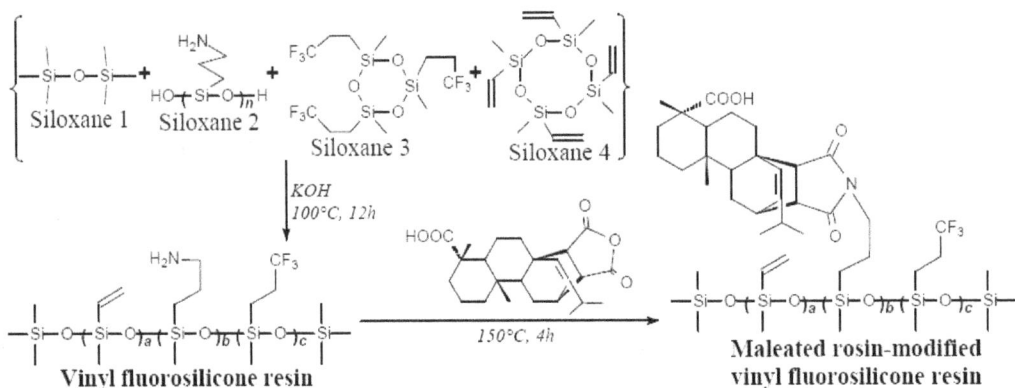

Scheme 73. Preparation of maleated rosin-modified vinyl fluorosilicone resin

Polysaccharides

Super hydrophobic wood can be prepared using maleated rosin, aluminum chloride and tetrabutyltitanate according to Scheme 74. Such a hydrophobic material is appreciated in construction. Related starch-rosin materials have been also prepared.

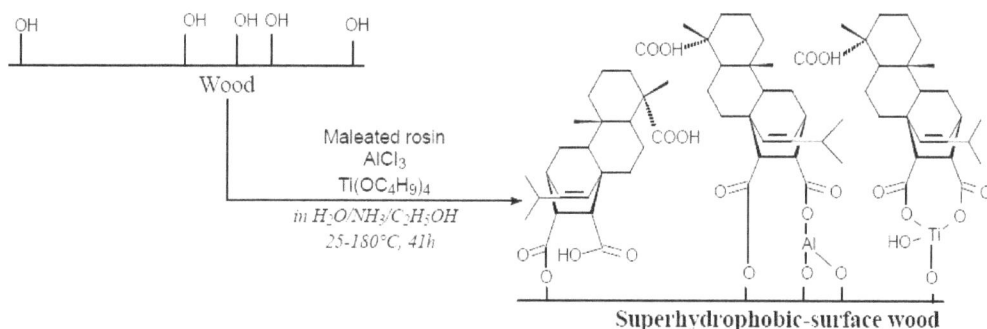

Scheme 74. Preparation of rosin-impregnated super hydrophobic wood

Rosin-esterified starch biopolymer can be prepared according to Scheme 75, prior to filtration, precipitation, washing and drying. Rosin decreases solubility and swelling of starch, which potentially allows using this polymer in food and biomedical materials.

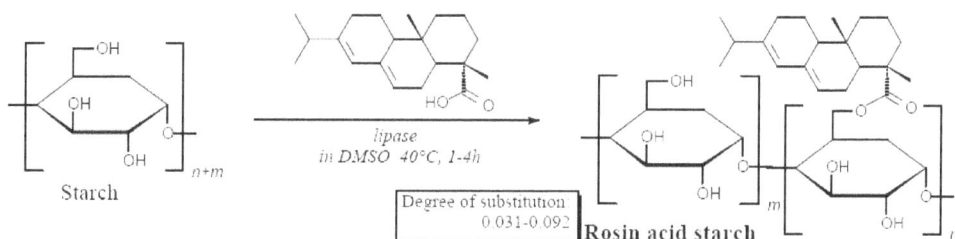

Scheme 75. Preparation of rosin-esterified starch

Cellulose nanofibers and nanocrystals surface-modified by rosin can be prepared according to Scheme 76, prior to washing in ethanol. They can be used as antibacterial reinforcements in bio-based package films. In similar way hemp fibers can be modified by tall oil rosin acids in order to improve the reinforcement adhesion to an epoxy matrix.

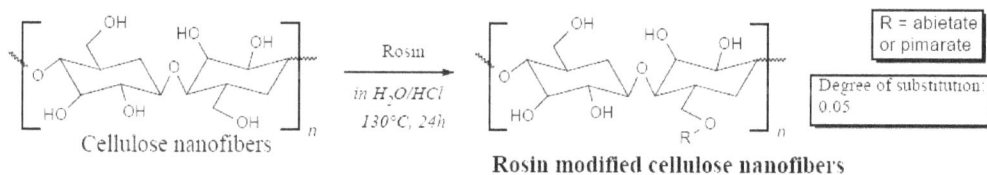

Scheme 76. Preparation of rosin-modified cellulose nanofibers

Chitosan grafted by rosin acrylate is a solid. It can be prepared from chitosan and rosin acid hydroxyethyl acrylate according to Scheme 77 prior to precipitation, filtration, washing and drying. It can be potentially used in controlled release applications.

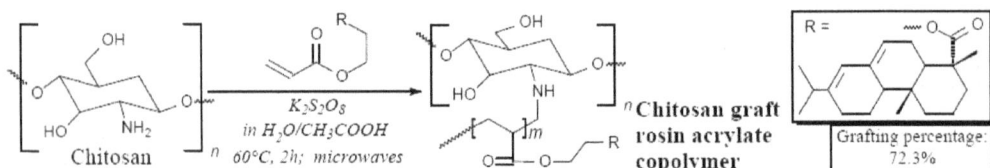

Scheme 77. Preparation of chitosan graft rosin acrylate copolymer

Other Materials

Dehydroabietic acid-glycerol carbonate product can be prepared from dehydroabietic acid and glycerin carbonates according to Scheme 78. It can be applied in xerographic tonners.

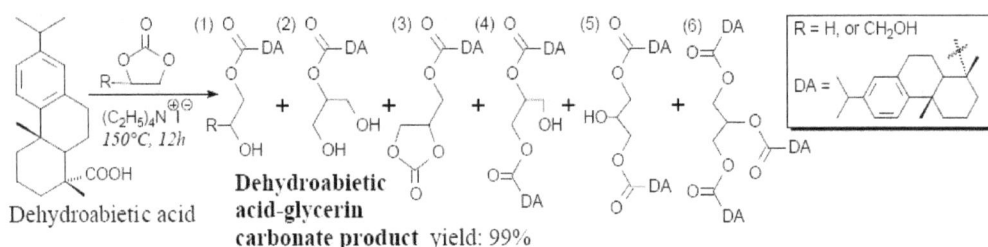

Scheme 78. Preparation of rosin-glycerin carbonate xerographic toner

Rosin-polyphthalate resin is a glossy light yellow solid. It can be prepared from maleopimaric acid and phthalic anhydride-glycerol polyester according to Scheme 79 prior to vacuum drying. Its application is an environmentally-friendly phenol-free printing ink.

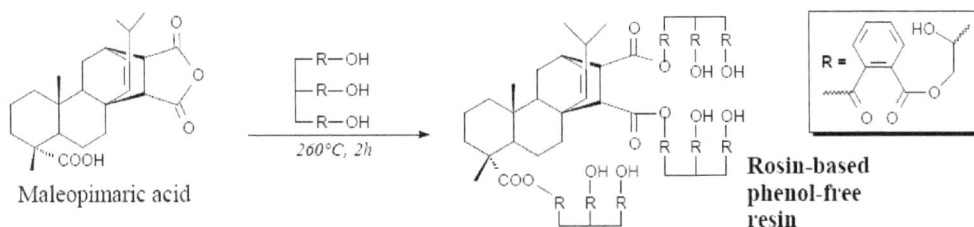

Scheme 79. Preparation of rosin-based phenol-free resin

Rosin propargyl ester is a white, transparent liquid. It can be synthesized from rosin chloride and propargyl alcohol according to Scheme 80, prior to washing and column chromatography. It can be used to prepare caprolactone graft copolymers.

Rosin ester containing poly(α-azide-ε-caprolactone) can be prepared from rosin propargyl ester, α-chloro-ε-caprolactone and sodium azide according to Scheme 80, and using such separation methods as centrifugation and vacuum drying. This biodegradable copolymer shows properties similar to poly(methyl acrylate). Similar polymers with quaternary ammonium groups showing antimicrobial properties can be also prepared.

Acrylic rosin methacryl diester can be prepared from acrylic rosin and 2-hydroxyethyl methacrylate according to Scheme 81, prior to rotary evaporation and vacuum drying. It can be used as acrylic monomer or resin.

Poly(styrene-co-acrylic rosin ester) can be prepared via suspension polymerization of acrylic rosin methacryl ester and styrene according to Scheme 81, prior to washing, filtration and vacuum drying. Its application is microspheres.

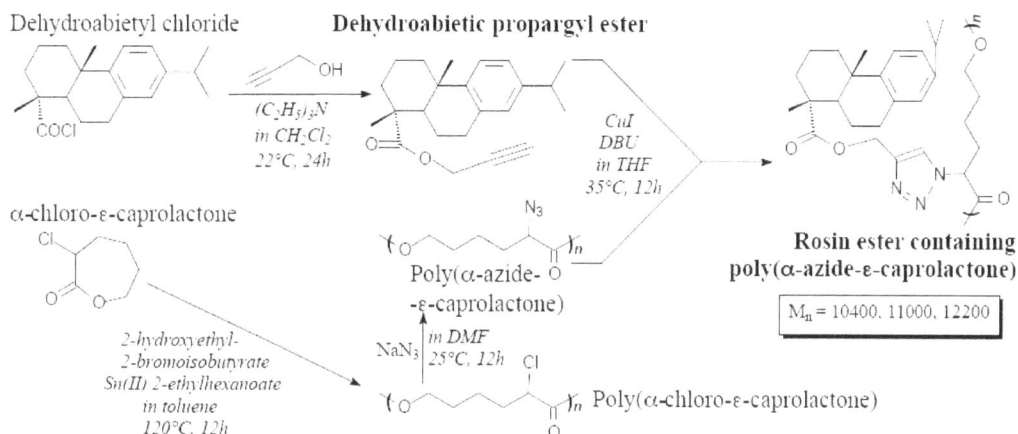

Scheme 80. Preparation of rosin ester containing poly(α-azide-ϵ-caprolactone)

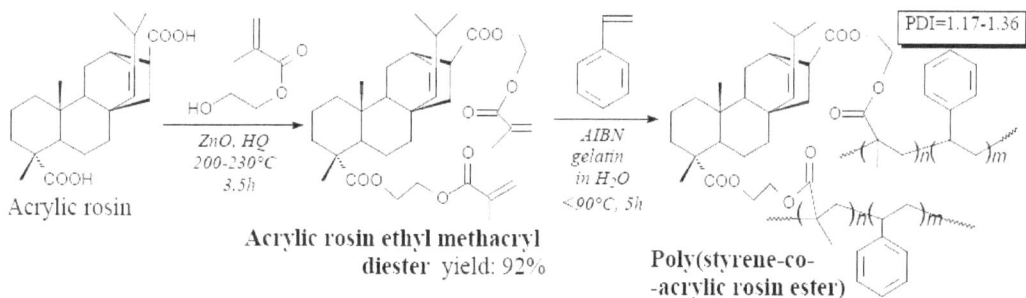

Scheme 81. Preparation of styrene and acrylic rosin ester copolymer.

Rosin-tung oil Diels-Alder adduct is a yellowish solid. It can be prepared from levopimaric acid and tung oil according to Scheme 82, prior to precipitation and drying. Its application is a filler, tackifier and adhesion promoter in polyurethane and UV-curable adhesives.

Acrylic-vinyl copolymer with rosin moieties is a slightly yellow solid. It can be prepared from rosin ethyl methacrylate, styrene and divinylbenzene according to Scheme 83, prior to washing, filtration and drying. Its application is polymer microspheres for adsorption and separation.

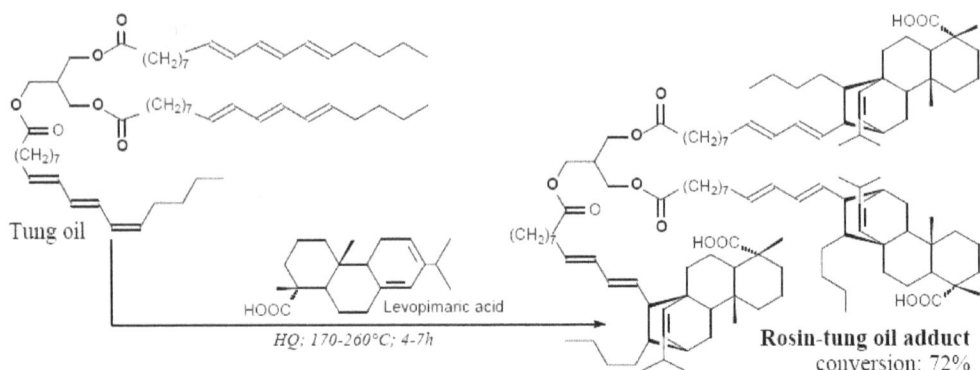

Scheme 82. Preparation of idealized rosin-tung oil Diels-Alder adduct

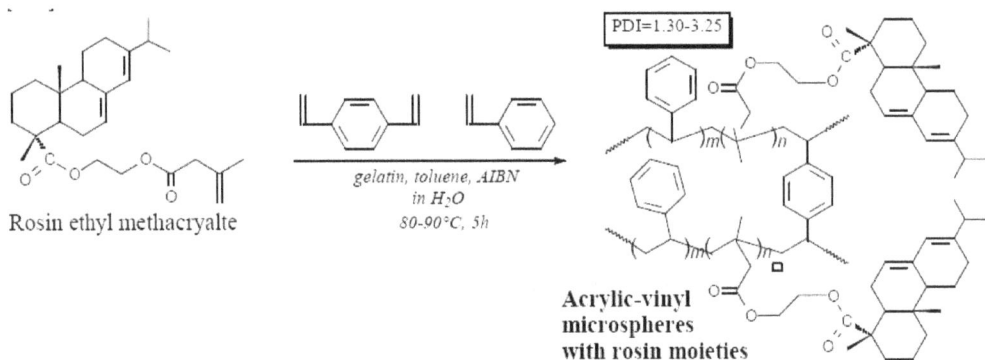

Scheme 83. Preparation of acrylic-vinyl copolymer with rosin moieties

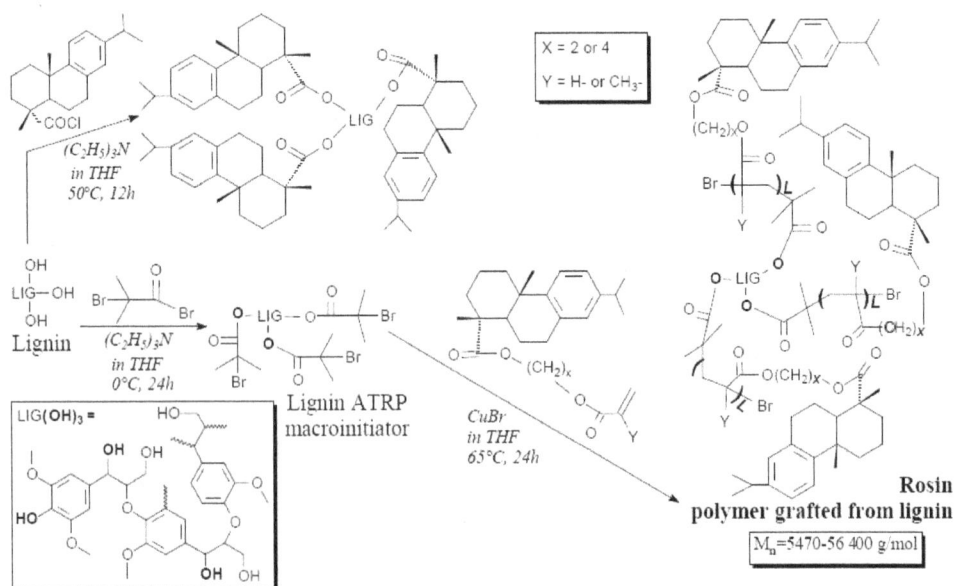

Scheme 84. Preparation of lignin materials hydrophobized by rosin

Rosin-caprolactone diblock copolymers can be prepared from dehydroabietic ethyl methacrylate, ε-caprolactone and 2-hydroxyethyl 2-bromoisobutyrate according to Scheme 85 prior to precipitation and washing. They can be applied in areas designed for simultaneously biobased and biodegradable materials.

Acrylpimaric acid hydroxyethyl polyesters are light yellow solids. They can be prepared via polycondensation of corresponding acrylpimaric diols according to Scheme 86 prior to grinding, washing and drying. Such polymers can be potentially applied in the modern electrical and electronic industries, especially for the environmentally friendly green products.

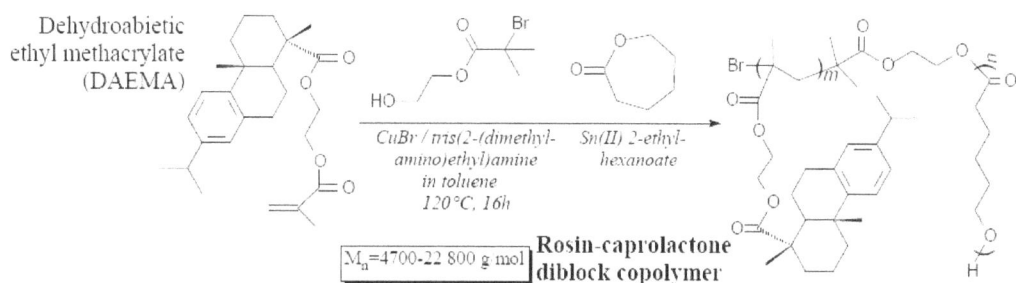

Scheme 85. Preparation of rosin-caprolactone diblock copolym

Furthermore, rosin can be pyrolized in order to prepare a matrix for silver nanoparticles to apply as antibacterial filler for wooden furniture or air filter for indoors, a catalyst carrier with potential application as counter electrode for dye-sensitized solar cells, a coating for bentonite particles and support for Fe2O3 nanoparticles for chromium ions adsorption, and activated carbons.

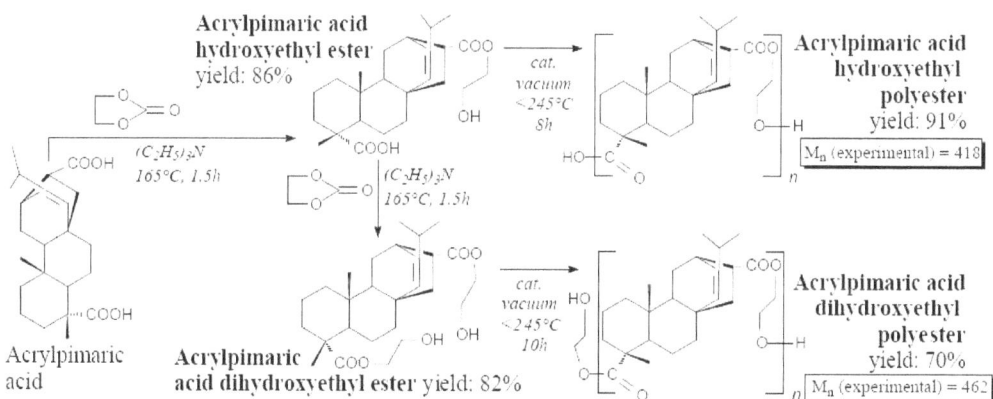

Scheme 86. Preparation of acrylpimaric acid hydroxyethyl polyester

CONCLUSION

Rosin is a highly modifiable raw material for both low molecular weight products and polymers. Its natural origin is accompanied with low price and a diterpene chemical nature, which is ready to introduce useful reactive chemical groups, and it exhibits a stiff constitution improving many thermal, physical, mechanical and functional properties of the final materials. In recent years rosin based chemicals have attracted growing interest. They offer great opportunities to produce useful products: resins, curing agents, surfactants, medicines, biocides, and materials for biomedical application, elastomers, coatings, adhesives, sorbents and catalysts.

Taking into consideration the declared properties of the prepared chemicals they seem to be competitive alternatives to existing products on the market. According to our subjective opinion, the following information should be provided (or obvious) for a recipe to be considered as complete: product name, product morphology, substrate(s) names, reaction scheme, catalyst, media, and temperature, pressure, time, yield and separation techniques.. However, this does not change the fact that rosin chemistry is able to deliver a serious amount of new environmentally-friendly solutions in many fields of science, medicine and engineering. The current review confirms this, and encourages further intensive research on rosin in the near future.

25
Chapter

SHELLAC BASED POLYMER

INTRODUCTION

Coated Dosage Forms

Film coating has become a routine operation in the production of solid oral dosage forms. There are numerous reasons for the application of film coatings to drug formulations. Polymeric film coatings can be applied to pharmaceutical solids for decorative purposes to provide gloss. The incorporation of dyes and pigments into the coating films allows coloration of the dosage form e.g. to facilitate product differentiation. Film coatings are also applied to improve the mechanical stability and reduce abrasion of the dosage form during manufacturing, shipping and storage. Sensitive drug formulations are film-coated to protect the ingredients from light or humidity. Film coatings can mask unpleasant taste as well as odor and facilitate swallowing which will improve the patient's compliance during drug therapy. Another important function of film coatings manifests itself once the dosage form reaches the GI tract. By selection of specific coating materials the film coating allows controlled release of the drug in the GI tract: So-called enteric coatings are applied to solid oral dosage forms to protect the drug from the acidic milieu of the stomach or vice versa. Sustained release coatings deliver the drug over a long time period and allow a reduction of daily intake frequency and thus improve the compliance of the patient. Especially for such coated controlled release formulations a consistent quality of the film coating is essential to maintain reproducible release profiles and to avoid the risk of dose dumping or loss of efficacy. The large number of applications for film coatings, the variety of coating materials and the need for reliable coating processes explains the great research effort on coated dosage forms.

FILM COATING MATERIALS

For each desired application a variety of different coating materials with tailor-made physicochemical properties is available. Most commonly used are polymers such as polymethacrylates, povidones and cellulose esters and ethers. Povidones and polymethacrylates are synthetic polymers which are obtained by emulsion polymerisation. The properties of the final polymer can be adjusted by selection of specific monomers. Cellulose derivatives are semi synthetic polymers gained either by esterification or etherification of natural cellulose. The type of substituent and the degree of substitution defines the properties of the final polymer. Besides these major films coating polymers there are few other materials used for film coating applications. One of them is shellac. The chemical properties of the coating material define the functionality of the coating film and thus the drug release characteristics of the final dosage form. Film coatings for taste masking or moisture protection are usually not intended to modify drug release. These coatings should maintain their barrier function during storage as well as during intake of the dosage form. Once the formulation reaches the stomach the coating should dissolve rapidly and release the drug. This type of coating is usually prepared with water soluble polymers but also with water insoluble, basic polymers that dissolve in the acidic milieu of the stomach. However, also the application of thin layers of enteric coatings has been approved for this purpose. Especially in moisture protective coatings the addition of pigments such as titanium dioxide or talc to the coating film further decreases the permeability of water vapor and thus enhances moisture protection Enteric coatings are applied to solid oral dosage forms to improve the chemical stability of acid-sensitive drugs , to decrease gastric irritation and to target the drug to the colon . Enteric coatings remain intact as long as the pH is below the release pH, above which the drug is released. For this application generally acidic polymers are used. They are protonated and thus insoluble in the acidic environment stomach but are permeable or dissolve at a higher pH. For the application of sustained release coatings usually water insoluble polymers are used. After swelling of the coating film or dissolution of incorporated pore formers the coating film becomes permeable and/or the drug release occurs by slow diffusion of the drug through the coating layer.

Film Formation

Water soluble coating materials are usually applied from aqueous solutions. However, many coating polymers, especially those for modified release applications, are water insoluble and cannot be applied from aqueous solutions. This is critical because the use of organic polymer solutions for the coating of pharmaceutical dosage forms has several disadvantages such as regulatory requirements, explosion hazard and limits for solvent residues in the final product. Hence, during the last decades aqueous coating systems have gained importance. Water insoluble polymers may be applied from aqueous polymer dispersions. These coating systems are either prepared by emulsion polymerization of monomers or by emulsification of a preformed polymer resulting in latex or pseudolatex formulations, respectively. In contrast to aqueous solutions in aqueous dispersions the polymer is colloidally dispersed in water. Aqueous dispersions show comparably low viscosities even at high polymer concentrations. However, polymer dispersions are sensitive to high electrolyte concentrations, pH changes, and high shear forces. There is a fundamental difference in the film forming mechanism between polymer solutions and aqueous dispersions. Film formation from polymer solutions results from the evaporation of the solvent, which leads to an increase in the polymer concentration and to an inter diffusion of the polymeric chains (Fig. 1). At higher polymer concentrations, an intermediate gel-like stage is reached. Upon further evaporation of the solvent, a solid polymeric film is obtained.

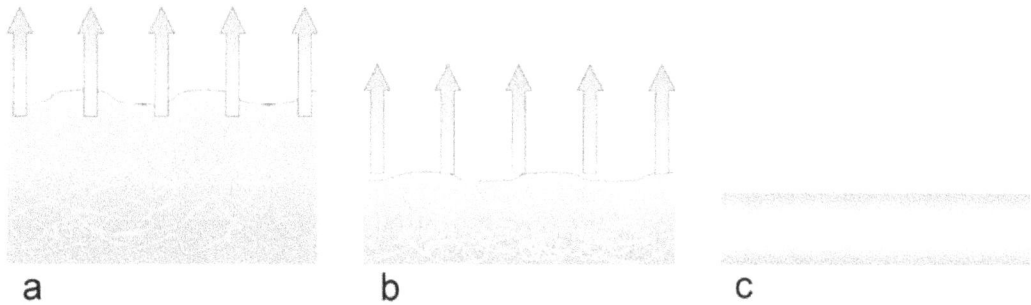

Fig. 1: Film formation from polymer solutions: solvent evaporation; b) intermediate gel-like stage; c) solid film

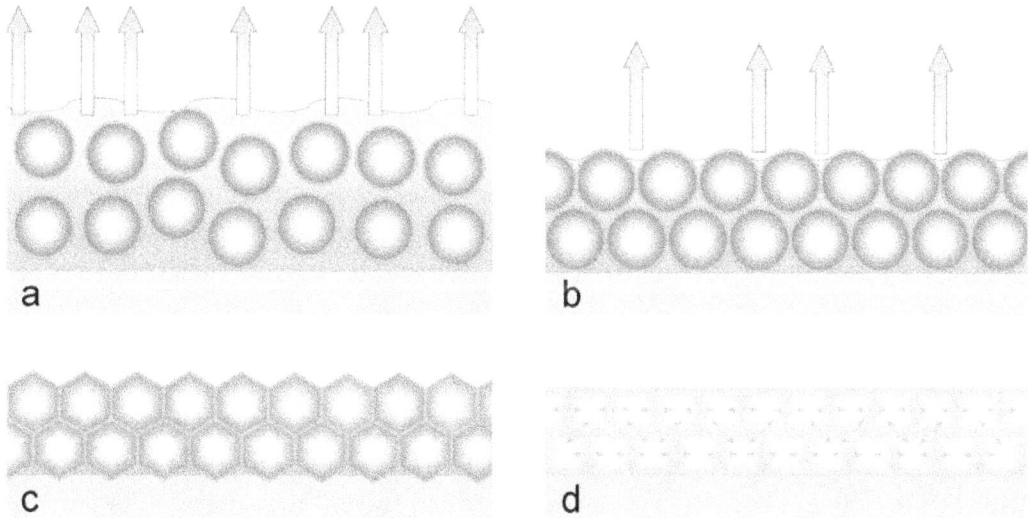

Fig. 2: Film formation from aqueous dispersions: a) solvent evaporation; b) close-packed arrangement; c) deformation of latex particles d) coalescence of latex particles above the MFT

Film formation from aqueous polymer dispersions is more complex (Fig. 2). As with the aqueous solutions, film formation starts with the evaporation of water from the coating formulation. In contrast to solutions, in aqueous dispersions latex particles are colloidally dispersed in water. As water evaporates the dispersed latex particles come into contact with each other and are forced into a closely packed, ordered array with water-filled voids. Further evaporation of water leads to particle deformation due to capillary pressure and interfacial tension and finally to coalescence of the particles. The film formation is completed by inter-diffusion of polymer chains through the particle interfaces. This coalescence of the latex particles occurs only above a minimum film forming temperature. Below this temperature the particles do not coalesce resulting in a failure in film formation.

FILM COATING PROCESS

Generally, there are three different techniques for the application of film coatings to pharmaceutical dosage forms: Pan coating, top spray fluid bed coating and bottom spray fluid bed coating which is also known as Wurster-based fluid bed coating. Because in the present work all coating experiments were performed in a Wurster coater, this process will be explained in more detail. In the fluid bed process agitation, turbation and drying of the product is achieved by the inlet air flow. In all modern coating processes the coating formulation is applied through a spraying nozzle. Whereas this spraying nozzle is located above the fluid bed in the top spray arrangement it is positioned on the bottom of the product chamber in the bottom spray arrangement, in the so-called Wurster insert. In contrast to the top spray arrangement, this configuration allows a well controlled product movement in a circulating fluid bed.

The product flow in the Wurster process is schematically displayed in Fig. 3. The product is pneumatically accelerated towards the spraying nozzle by atomizing air and passes the spraying nozzle where the coating formulation is applied. Subsequently, the product leaves the Wurster insert, slows down and finally drops back into the fluid bed where the product dries and the circulation starts again. The coating process consists of three phases. During the start up phase the inlet air preheats product and equipment. This heating prevents over wetting and facilitates film formation during the initial application of the coating formulation. During the coating phase the coating formulation is applied to the dosage form. The atomizing air transports droplets of the coating formulation from the spraying nozzle to the passing substrate. The droplets spread on the surface, dry by the inlet air and form the coating film. In the final drying phase residual solvent is evaporated to prevent sticking of the product. After that, the product is cooled down and equilibrated to ambient conditions.

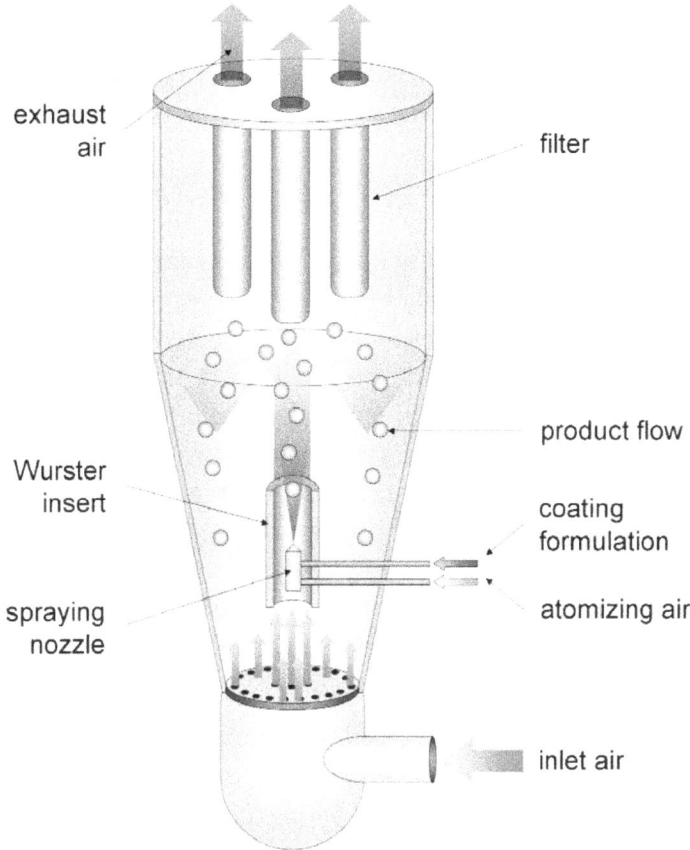

Fig. 3: Schematic description of the Wurster coating process

There are many parameters that affect the quality of the coated product. The diameter and length of the Wurster insert as well as its distance from the bottom plate influence the product movement. Inlet air pressure and atomizing air pressure must be balanced and adjusted to the coater's dimensions to assure an adequate product movement. The temperature of the inlet air defines its drying capacity and thus its moisture content. Nozzle diameter, spray rate and atomizing air pressure regulate the droplet size of the coating formulation. Hence, these parameters have to be related to each other as well as to the inlet air temperature and volume to avoid spray drying of the coating formulation on the one hand and over wetting and agglomeration of the product on the other hand. Due to this complexity much effort is directed towards the optimization of the process parameters.

Shellac

Shellac is a natural product with interesting properties and an exceptional versatility. Shellac is the purified product of the natural resin lac which is the hardened secretion of the small, parasitic insect Kerria Lacca, popularly known as the lac insect. It is the only known commercial resin of animal origin. Lac has been known in India and China since ancient times. Its use can be traced back to recordings from India from more than 2000 years ago. The first mentioning in Europe can be referred to van Linschoten in 1596 who was sent to India by the king of Portugal.

ORIGIN

The lac insect belongs to the family Kerriidae and superfamily Coccidae to which also the scale insects and the mealy bugs belong. The geographical distribution of this family is very wide and species have been collected from all continents except Europe. However, despite this wide distribution the main production of shellac takes place in South-eastern Asia especially India, Thailand and Myanmar. A number of species of the genus Kerria is known. However, Kerria Lacca is by far the most important species, producing the major percentage of commercial lac. The lac insect lives on certain trees and bushes, the so-called lac hosts. Even though many of these lac hosts exist, only few are used for large-scale cultivation. The major hosts in India are the Palas tree (Butea monosperma), the Ber tree (Zizyphus mauritiana) and the Kusum tree (Schleichera oleosa). In Thailand the major host is the Rain tree (Samanea saman).

Young larvae of lac insects are red and measure about half a millimetre in length and half as much in width. After emergence they settle down on the lac host and attach themselves to the host by piercing its bark. They suck the sap of the host and start secreting lac. Under this coating the larvae grow while they continue the secretion of lac from the inside. After eight to fourteen weeks the male insect emerges out of its lac cover, fertilizes the female and dies soon after. The female continues growing and increases lac secretion until the egg laying period. There are two generations of lac insects and thus at least two crops per year. The crops are collected by cutting down the lac-bearing twigs and scraping off the so called stick lac. The yield and quality of lac may vary considerably depending on the insect species and the lac host. As the quality of the final product shellac is directly dependent on the type of raw material, this great variety and the lack of cultivation can be critical for production of shellac with a consistent quality.

REFINING PROCESS

The harvested sticklac is cleaned from wood and insect residues. A subsequent washing step with water removes soluble ingredients (e.g. laccaic acid) and leads to the intermediate product Seed lac. Its color can vary from pale yellow to deep red depending on the insect strain and host tree.

There are three different ways of refining seed lac, resulting in different shellac qualities. Wax-containing shellac is obtained by the traditional melting filtration process where molten seed lac is pressed through a filter and cast to a film. After cooling the film breaks into the typical flakes.

a   b

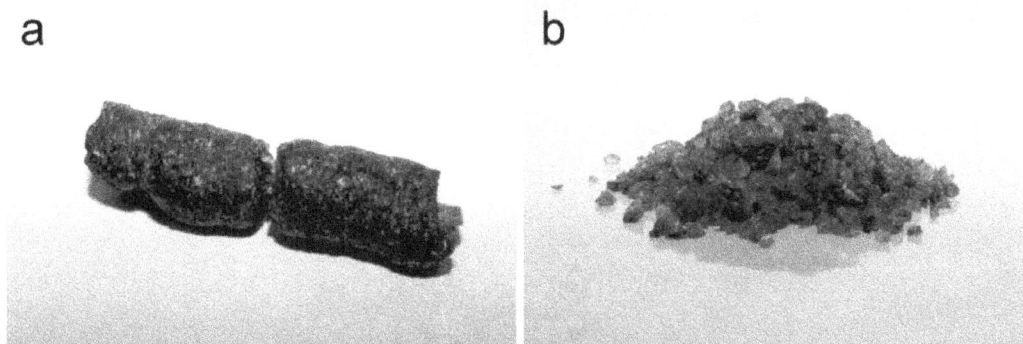

Fig. 4: a) raw material sticklac; b) intermediate product seedlac

This kind of shellac contains shellac wax and its color directly corresponds to that of the seed lac used. The color of seed lac is mainly attributed to the presence of the dye erythrolaccin. To obtain colorless shellac this dye either has to be eliminated or to be bleached. Bleached shellac is gained by dissolution of seed lac in aqueous alkali solutions followed by treatment with sodium hypochlorite. Shellac is then precipitated by addition of sulphuric acid. Solutions of bleached shellac are almost colorless which is advantageous for many applications. However, the bleaching process leads to changes in the molecular structure such as chlorination resulting in a higher reactivity and thus reduced stability. Shellac obtained by melting or bleaching processes is usually intended for technical use.

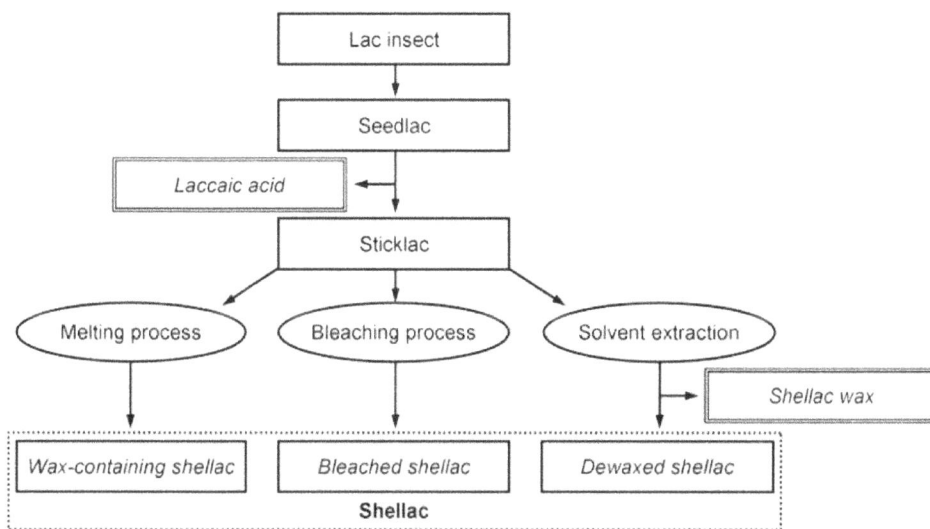

Fig. 5: Flow chart of the refining process of shellac

Shellac for pharmaceutical applications is usually de-waxed shellac that is refined by solvent extraction: First, seed lac is dissolved in ethanol. Impurities and shellac wax are removed by filtration. Subsequently, the shellac solution is decolorized by addition of activated carbon. After removal of the activated carbon by a second filtration, the solvent is evaporated in a thin film evaporator and recovered. Removal of the solvent increases the concentration of the shellac solution until a hot molten shellac mass is obtained which is cast to a film. After cooling the film breaks into shellac flakes. Solvent extraction is a gentle process that does not affect the molecular structure. It allows the production of shellac with narrow specifications.

Fig. 6: Solvent extraction: a) tank with ethanolic shellac solution; b) discharge of molten shellac from a thin film evaporator; c) cooling of cast shellac film; d) shellac flakes

COMPOSITION OF SHELLAC

Early studies on the chemical composition of lac can be referred to Hatchett in 1804. Since this time many researchers from all over the world performed studies on the composition of this complex material. Shellac is a natural material with a complex mixture of esters and polyesters of polyhydroxy acids. The first systematic analysis of its composition was performed by Tschirch et al. in 1899 after fractionation of the material in different solvents.

Variations of this method have been used up to the present for separation of the shellac components. The molecular structure of the ingredients was analyzed and revised several times until the structure of the main components aleuritic acid and shellolic acid was clarified.

It was found that depending on the shellac type aleuritic acid and homologues of shellolic acid make about 70 percent of the total shellac composition. In later studies butolic acid and other sesquiterpenic acids related to shellolic acid were identified as further components of the lac resin. Besides the individual acids also several esters as well as the position of the ester linkages have been identified. These findings have been confirmed and further specified by modern analytical methods such as liquid and gas chromatography or combined pyrolysis and mass spectrometry. However, in spite of all this attention, the composition of shellac is still not completely understood. A reason for this might be the fact that the composition of shellac is highly variable depending on its origin and the type of refining.

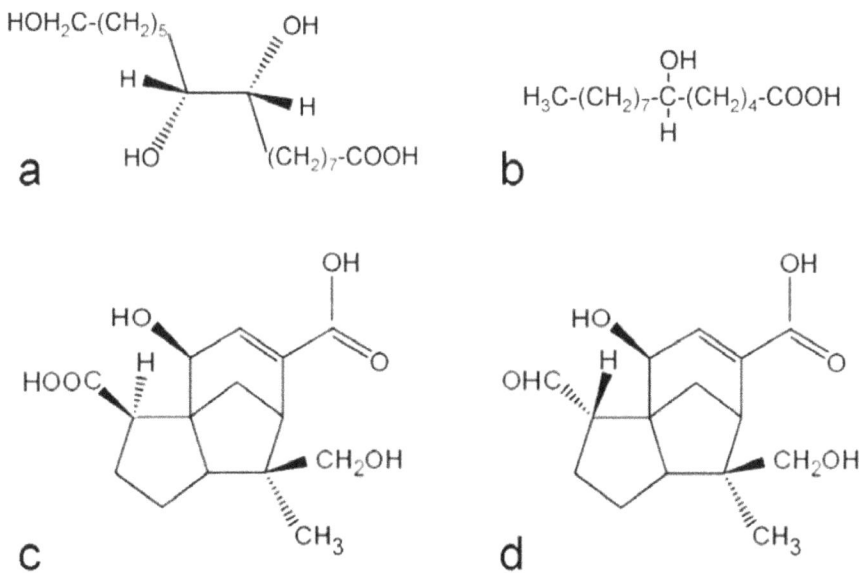

Fig. 7: Main components of shellac: a) aleuritic acid; b) butolic acid shellolic acid;

d) jalaric acid

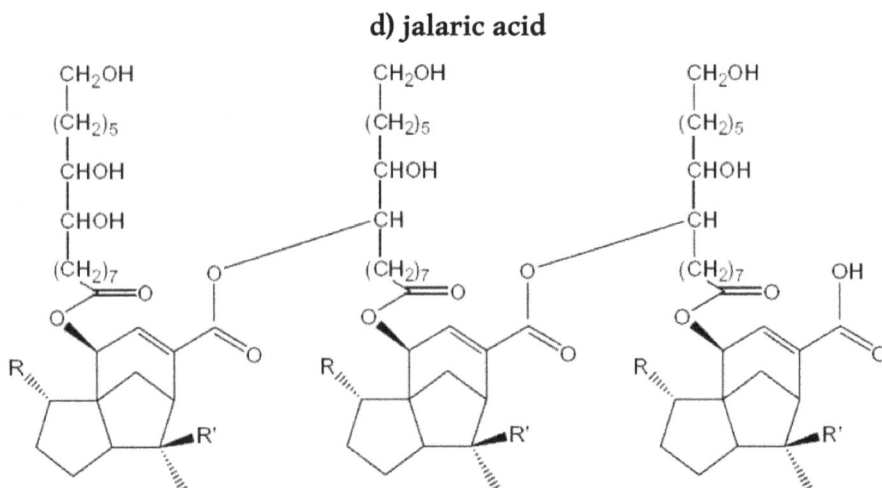

Fig. 8: Chemical structure of shellac according to Limmatvapirat et al

Properties of shellac

Shellac is a hard, brittle and resinous solid. It is practically odorless in the cold but evolves a characteristic smell on heating and melting. This smell can partially be referred to aleuritic acid which is known to be a starting material for the production of flavors. Its color is dependent on the type of seed lac and the refining process and can range from pale yellow to deep red. The color of the material is usually characterized by the Gardener or Lovibond scale. Shellac films provide high gloss, a low permeability for water vapor and gases and good dielectric behavior.

Shellac is water insoluble. However, by addition of alkali translucent aqueous solutions can be obtained. Shellac is soluble in ethanol, methanol and partially soluble in ether, ethyl acetate and chloroform. Even though few allergic reactions of the skin and the respiratory tract are reported for shellac-containing products the material is generally regarded as non-toxic and physiologically harmless. The pharmacopoeias characterize shellac only by the acid value. The Ph. Eur. allows a range of the acid value between 65 and 80. The acid value of most dewaxed shellac types is about 70. Whereas the acid value of wax-containing shellac or bleached shellac can be considerably higher, the acid value of aged shellac may be significantly lower. In contrast to crystalline substances the amorphous material shellac has no sharp softening or melting point. Its glass transition temperature depends on the shellac type and varies between 30 and 50°C for the acid form. It has been reported that glass transition temperatures of ammonium salts of shellac can be significantly higher.

Shellac undergoes aging. Since most of the acids contain more than one hydroxyl group and some more than one carboxyl group it is believed that this aging is a result of selfesterification of the material. This esterification is accompanied by a loss of solubility, a decrease in the acid value and an increase in the glass transition temperature. This aging manifests itself in the so-called blocking of the material as the individual shellac flakes stick together. Several investigations deal with the prevention of shellac aging. It has been reported that proper storing conditions at temperatures below 27°C and protection from light as well as the addition of antioxidants prolong stability. It turned out that the stability of shellac can greatly be improved by salt formation with ammonia or organic bases such as 2-amino-2-methyl-1-propanol. It is assumed that this salt formation leads to sterical hindrance and thus a reduced selfesterification.

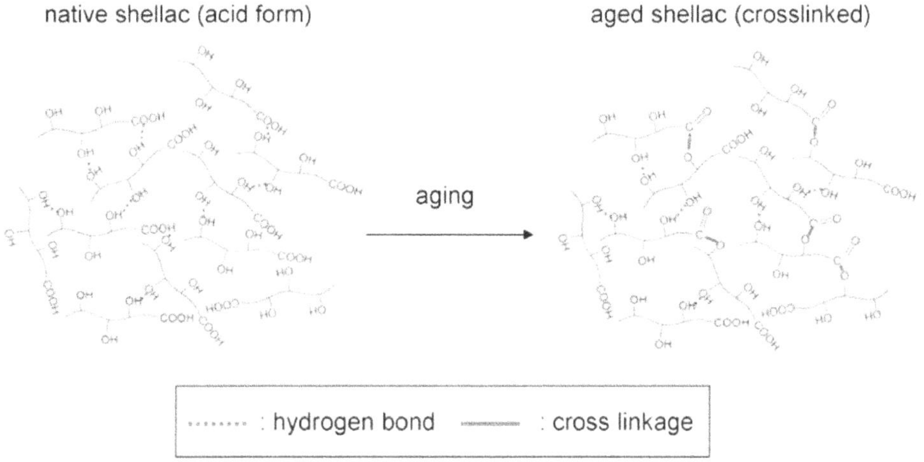

Fig. 9: Schematic description of aging of shellac; adapted from Limmatvapirat et al.

Modification of shellac

Apart from the use of different salt forms modifications of shellac itself have been discussed to obtain the desired material properties. Shellac was partially hydrolyzed to achieve a higher solubility. However, it turned out that partially hydrolyzed shellac was less stable. Esterification was performed either with organic acids to improve enteric film properties or with glycerol to obtain a better microencapsulating material. Shellac was treated with gamma radiation to enhance biodegradability; it was crosslinked with propanediamine and t-butylacetoacetate or graft polymerized with synthetic polymers to improve water resistance of shellac films or the dielectric behavior of the material.

Applications of shellac

The use of shellac as a binding material in music records may probably be the best known application of the material. Due to its brittleness shellac was soon substituted by synthetic polymers such as polyvinylchloride which shows a better mechanical stability. However, apart from music records there is still a great variety of applications for this versatile material. Some of these are listed in the following paragraphs:

Shellac has a long tradition as ingredient in colors and lacquers where it has been used as protective layer on artistic objects and music instruments. In dentistry shellac is used as a material for dental base plates and impression trays. Modified with fluoride and epoxy resins shellac has also been used as a varnish to reduce dental hypersensitivity.

Shellac is listed as GRAS (generally recognized as safe) by the FDA. This regulatory status allows its use as additive in food products which is the most common application of shellac. Shellac is used as coating material on confectionaries and fruits to apply gloss. Moreover, the application of shellac coatings on citrus fruits reduces the loss of water and volatile flavouring substances. It has been reported that shellac coatings prevent early penicillum-induced postharvest decay by supporting populations of bacterial and yeast antagonists. Also the application of shellac in the meat industry has been discussed. Cattle hides have been treated with shellac solutions during slaughter with the aim to reduce bacterial contamination of the meat .The application of shellac as matrix material in so-called biocomposites for the production of biodegradable composite materials may also be of interest as well as the application in paper-based packaging materials to improve water resistance.

PHARMACEUTICAL APPLICATIONS

Shellac coatings for food applications are commonly applied from ethanolic solutions. This is critical, since the organic coating process requires special equipment for prevention of explosion and removal as well as recovery of the solvent. Various formulations have been developed to replace organic shellac coating systems. Aqueous formulations were prepared as dispersions by high pressure homogenization or as pseudolatex. Another attempt was done with shellac as material for powder polymer coating, whereby the dosage forms are coated with micronized shellac powder and high amounts of plasticizer. However, as with the organic solutions all these alternatives contain shellac in its acid form, which undergoes aging. This aging is accompanied with a change of the physicochemical properties of the material and results in significant changes in the drug release profile of shellac-containing dosage forms. Because of this instability the use of shellac in pharmaceutical applications has declined. Since the introduction of aqueous ammoniacal solutions, shellac regained importance for pharmaceutical dosage forms. Aqueous shellac solutions are easy to handle and show a low viscosity even at high shellac concentrations. Besides these technical advantages, the coating films prepared from these solutions result in the ammonium salt, which lacks the instability problems of the acid form. Due to its acidic character shellac is mostly used as an enteric coating. However, shellac has a comparably high dissolution pH of about 7.3. This is unsuitable for a conventional enteric coating since it requires the addition of suitable additives to achieve fast release in the proximal small intestine. Organic acids and hydrophilic polymers have been added to act as pore formers or swelling agents to enhance drug release. Shellac was modified to improve its solubility at lower pH. Partial hydrolysis by alkali treatment resulted in a better solubility and improved mechanical stability of the shellac films. However, these films showed pronounced aging. Another approach was the esterification of shellac with succinic acid. This esterification with the dibasic carboxylic acid resulted in an increase in the acid value, which was accompanied by improved solubility. Whereas the high dissolution pH of shellac is unsuitable for a conventional enteric coating it is of interest for colon targeting formulations. The shellac coating layer remains intact during the passage of the stomach and the small intestine until it reaches the colon with its higher pH. This allows the transport of drugs into the colon for a topical treatment of colonic diseases. Moreover, the peptidase activity in the colon is lower than in the upper GI tract allowing for an oral delivery of peptide drugs such as insulin.

Shellac has also been used as matrix former in sustained release tablet and pellet formulations. It could be shown that drug release was prolonged depending on the drug/shellac ratio. This sustained release effect could be further improved by a subsequent thermal treatment at different temperatures. Its low permeability for water vapor and gases qualifies shellac as a moisture protective coating for water sensitive formulations. Moreover, its use as taste masking material has been discussed .Another application of shellac is microencapsulation. It has been used in its natural form as the encapsulation material itself, as additional coating on gelatin microspheres or modified by esterification with glycerol to improve the encapsulation properties. Microencapsulation has also been performed by precipitation of shellac with calcium ions to obtain water insoluble microspheres. This wide field of applications of this versatile material explains the great effort which is directed towards research on shellac.

REFERENCES

Crivello, J. V.; Narayan, R. Chem. Mater. 1992, 4, 692.

Lligadas, G.; Ronda, J. C.; Galia, M.; Cadiz, V. Biomacromolecules 2006, 7, 3521.

Park, S. J.; Jin, F. L.; Lee, J. R. Macromol. Rapid Commun. 2004, 25, 724.

Cakmakli, B.; Hazer, B.; Tekin, I. O.; Comert, F. B. Biomacromolecules 2005, 6, 1750.

Tsujimoto, T.; Uyama, H.; Kobayashi, S. Macromolecules 2004, 37, 1777.

Andjelkovic, D. D.; Larock, R. C. Biomacromolecules 2006, 7, 927.

Liu, Z. S.; Sharma, B. K.; Erhan, S. Z. Biomacromolecules 2007, 8, 233.

Gandini, A. Macromolecules 2008, 41, 9491.

Lligadas, G.; Ronda, J. C.; Galia, M.; Biermann, U.; Metzger, J. O. J. Polym. Sci., Part A: Polym. Chem. 2006, 44, 634.

Holland, J. M.; Lewis, M.; Nelson, A. J. Org. Chem. 2003, 68, 747.

Lligadas, G.; Ronda, J. C.; Galia, M.; Cadiz, V. Biomacromolecules 2007, 8, 686.

Sharma, B. K.; Adhvaryu, A.; Erhan, S. Z. J. Agric. Food Chem. 2006, 54, 9866.

Chakraborti, A. K.; Rudrawar, S.; Kondaskar, A. Eur. J. Org. Chem. 2004, 3597.

Schilling, P. U.S. Patent 4,597,799, 1986.

Ballard, R. L.; Tuman, S. J.; Fouquette, D. J.; Stegmiller, W.; Soucek, M. D. Chem. Mater. 1999, 11, 726.

INDEX

www.ingramcontent.com/pod-product-compliance
Lightning Source LLC
Chambersburg PA
CBHW082005190326
41458CB00010B/3079